Fruit Crops
Cultivation Practices and Economics

NIPA® GENX ELECTRONIC RESOURCES & SOLUTIONS P. LTD.
New Delhi-110 034

About the Authors

Parshant Bakshi obtained his B.Sc. Agri. (Hons.) from M.D.S. University, Ajmer in 1998, M.Sc. (Horticulture) from Dr. B.R. Ambedkar University, Agra in 2000 and Ph. D. (Pomology & PHT) from Sher-e-Kashmir University of Agricultural Sciences and Technology of Jammu, Chatha, Jammu, Jammu & Kashmir in 2004. He has been awarded MASHAV fellowship by the Ministry of External Affairs, Israel, in 2012 and has been adjudged for best poster and oral presentation at various national and International platforms. He is having more than 13 years of experience in teaching, research and extension activities. He is having more than 100 publications in national and international journals of repute. He has guided 5 research students (M.Sc. and Ph.D.) as chairman and 25 as co-chairman /member and authored 7 books and 9 manuals. He handled 03 externally funded project as P.I. and 14 as Co-P.I. He remained Head, Advanced Centre for Horticulture Research, Udheywalla of Sher-e-Kashmir University of Agricultural Sciences and Technology of Jammu, Chatha, Jammu, Jammu & Kashmir from 2016-2018. He is the life member of 12 societies of Horticulture and editorial board of 7 professional journals. He is the fellow of Hi-Tech Horticultural Society. He delivered 20 T.V and radio talks and has RG score of 20.21. His area of research includes; enzymological changes in fruits, maturity indices of sub-tropical fruits, foliar application of micro-nutrients, rejuvenation of old/senile orchards, high density planting and molecular characterization of mango and walnut.

Anil Bhat, Assistant Professor, Agricultural Economics, Division of Agricultural Economics & Statistics, SKUAST-Jammu possesses a meritorious academic record. He has done M.Sc. (Agril. Economics) from the Chaudhary Charan Singh University, Meerut and Ph.D. (Agril. Economics) from the Sher-e-Kashmir University of Agricultural Sciences and Technology of Jammu. He also holds the Post Graduate Diploma in Rural Development from Indira Gandhi National Open University, New Delhi and Post Graduate Diploma in Agricultural Marketing from Pondicherry University, Kalapet, Puducherry. He has 40 publications including 12 book chapters and four books to his credit. He has guided 5 research students (M. Sc. and MBA Agri Business Management) as Major Advisor. He is the member of four professional societies and is also reviewer and editor of professional journals. Besides, he is also the recipient of Young Associate Scientist Award in the field of Agricultural Economics and has also received number of research paper presentation awards. His field of expertise is production, marketing and agri-business management.

Deep Ji Bhat, presently working as Assistant Professor in the Division of Fruit Science, Faculty of Agriculture, SKUAST-Jammu. He is in the profession of teaching, research and extension with experience of more than 09 years. He has handled 02 externally funded projects. He has guided 03 M.Sc. students and presently guiding 01 Ph.D student. He has published 25 research papers in various national and international journals. He has been recipient of outstanding professional award by Indian Society of Hill Agriculture in 2011. He is founder member of Journal of Hill Agriculture and life member of several professional societies.

Naveen Kumar Patle, working as Deputy Commissioner (Horticulture) with Ministry of Agriculture & Farmers Welfare, Krishi Bhawan, New Delhi, is possessing 20 years experience in Extension, Teaching, Research and Administration with different organizations at various capacities. Presently, he is involved in planning, policy making & implementation of various centrally sponsored programmes like Mission for Integrated Development of Horticulture and Rashtriya Krishi Vikas Yojna etc. for holistic development of horticulture in the country. He has published more than 25 Research Papers in various national & international journals and written popular articles/ book chapters in horticulture and agro forestry field. He has also undergone various national and international training on horticulture aspects. He is also a life member of various professional societies and representing different committees constituted by Ministry of Agriculture & Farmers Welfare.

Fruit Crops
Cultivation Practices and Economics

Parshant Bakshi
Anil Bhat
Deep Ji Bhat
Naveen Kumar Patle

NIPA® GENX ELECTRONIC RESOURCES & SOLUTIONS P. LTD.
New Delhi-110 034

NIPA® GENX ELECTRONIC RESOURCES & SOLUTIONS P. LTD.

101,103, Vikas Surya Plaza, CU Block
L.S.C. Market, Pitam Pura, New Delhi-110 034
Ph : +91 11 27341616, 27341717, 27341718
E-mail: newindiapublishingagency@gmail.com
Web: www.nipabooks.com

For customer assistance, please contact
Phone: + 91-11-27 34 17 17
Fax: + 91-11- 27 34 16 16
E-Mail: feedbacks@nipabooks.com

ISBN: 978-81-19103-90-4

Composed and Designed by NIPA®.

Government of India
Ministry of Agriculture and Farmers Welfare
Department of Agriculture, Cooperation
and Farmers Welfare, Krishi Bhawan
New Delhi-110001

Dr. B.N. Srinivasa Murthy
Horticulture Commissioner

Dated: 13 February 2019

Foreword

There has been a record production of Horticulture Crops with production during the year 2017-18(Final) reaching 311.7 Million Tonnes which is 3.7% higher than the previous year and 10% higher than the past 5 years' average production. Production of fruits is estimated at 97.35 million tonnes which is 4.8% higher than previous year. These production increases partly reflect rising demand, but there has also been a strong push from improved varieties and advances in fruit tree husbandry.India has emerged as second largest producer of fruits in the world. It has diverse climates and soil for cultivation of fruit crops providing ample opportunities for the development of fruit industry but the greatest challenge in present is to produce sufficient fruit to feed the ever-increasing human population.Horticulture is developing fast in more countries and so in India than the government can manage with; but within its controls the government should promote growth in horticultural economics at regional and state level. Horticulture sector in diverse forms is finding its way into everyday life in developed as well as in developing countries. Thus, the field of horticultural economics will in future be approached from several directions and its scope must surely be widened by practice more than by research.

The Jammu and Kashmir State is bestowed with climatic diversities having subtropical, intermediate and temperate zones and is best suited for growing various fruit crops. There is appreciable change in the concept of horticulture development in the state which leads to direct or indirect involvement of 7 lakh families with horticulture comprising of about 33 lakh people. Horticulture development is one of the thrust areas of state Government. A number of programmes have been implemented in the past, resulting in the generation of higher incomes in the rural areas, thereby improving the quality of life in villages.Fruit production of the state increased from 16.36 lakh MTs in 2007-08 to 23.92 lakh MTs in 2018-19, as per first advance estimate of the Ministry. The Jammu region of J&K state is emerging as fast growing horticulture sector with its contribution of about 7-8 per cent to the state's exchequer. Keeping in view its huge potential in terms of production, variety, quality and relative returns, this sector is being treated as one of the core sectors of the state's economy. Notwithstanding significant advances in some subtropical and intermediate fruits; the opportunities to grow, consume and export more fruit from this region remains under exploited due to improper economic analysis.

This book describes the complete cultivation, management and economic analysis of important fruits *viz.*, mango, kinnow, guava, litchi, apple, peach, pear, plum, pecan etc. The authors haveput inremarkable efforts to undertake important fruit crops in terms of cultivation practices and economic analysis. The cultivation practices, management, guidelines and cost and return analysis in the book is novel and are the platform for the author's detailed vertical descriptions. The information on fruit growing practices in this publication along with economic analysis applicable to all the areas of the state and a benchmark for the country.

I am assertive that the information contained with respect to income in this book will engender a greater fruit growing culture among farmers and will help them to increase their income by way of following good management practices and judicious use of inputs. I congratulate the authors for bringing out this book and hope thisbook will prove to be of immense value to our orchardists, students, extension workers, development officials, scientists and policy makers.

Dr. B.N. Srinivasa Murthy

Preface

Prosperity through fruit cultivation is the new slogan gaining much acceptance among policy makers. Presently, fruit culture has established its credibility in improving income through increased productivity, generating employment and enhancing export. The focused attention on investment in fruit production during the last decade has been rewarding in terms of increased production and productivity of fruits. India has emerged as second largest producer of fruits in the world. It has diverse climates and soil for cultivation of fruit crops providing ample opportunities for the development of fruit industry but the greatest challenge in present is to produce sufficient fruit to feed the ever-increasing human population. This demands infusion of technology for an efficient utilization of resources for deriving higher output per unit of input with excellent quality of the produce within short span of time. In the present era of open economy, it has become increasingly necessary that our produce is competitive, both in the domestic as well as in international markets. This would be possible only through deployment of high-tech application and precision farming methods. Fruits are very important for human beings and also play an important role in religious practices, mythology and art. They are not only delicious but also have many nutrients which are necessary for human health. Fruit production requires a lot of science and some basic fundamental knowledge to grow them successfully. This book compiles many fundamental issues of fruit production like layout and planting, cultural practices, growth and bearing habits of fruit crops, planting and propagation method, disease and pest management as well as cost of cultivation etc. Understanding of all these topics will help the students, fruit growers and policy makers for proper knowledge of fundamentals of fruit production as well as the investment in orchard establishment. We hope that this book would also be helpful for fruit growers, nurserymen, farmers, teachers, scientist, extension officer and all those who wish to become familiar with the topic in relation to their professional interest. We have tried to keep the language as simple and straight forward as possible and consistent with accurate representation of the content. Every effort has been made to present the ideas in very easy and understandable language and the interests of each reader. We would like to recommend "Fruit Crops: Cultivation Practices and Economics" to students, educators, researchers, extension specialist and orchard practitioners for cultivation practices of fruit crops and their economics.

We acknowledge our heart-felt gratitude to Prof. J. P. Sharma, Vice-Chancellor, Sher-e-Kashmir University of Agricultural Sciences and Technology of Jammu (SKUAST-Jammu) for his guidance, support and constructive suggestions in completing the book.

We are also thankful to New India Publishing Agency (NIPA), New Delhi for publishing this book. Finally, we would like to thank one and all who have contributed directly or indirectly in bringing out this book.

Parshant Bakshi
Anil Bhat
Deep Ji Bhat
Naveen Kumar Patle

Contents

1

Fruit Crops Cultivation and Management

In India, horticulture has emerged as one of the potential agricultural enterprise in accelerating the growth of economy. Its role in the country's nutritional security, poverty alleviation and employment generation programmes is becoming increasingly important. It offers not only a wide range of options to the farmers for crop diversification, but also provides ample scope for sustaining large number of Agro-industries which generate huge employment opportunities. At present, horticulture is contributing 24.5% of GDP from 8% land area. As a result of a number of thoughtful research, technological and policy initiatives and inputs, horticulture in India, today, has become a sustainable and viable venture for the small and marginal farmers. It is a matter of satisfaction that their food consumption levels and household income have increased. Besides, this sector has also started attracting entrepreneurs for taking up horticulture as a commercial venture. Therefore, there is a great scope for the horticulture industry to grow and flourish.

Jammu division with three agro-climatic zones *viz*., sub-tropical, intermediate (low hills) and temperate (Fig 1.1) has vast potential for horticultural industry by way of improving the productivity of existing orchards and using vast tracts of underutilized wastelands, kandi areas of sub-tropical zone and hilly areas of intermediate and temperate zones for fruit production. Horticulture is the back bone of the UT economy, providing livelihood to more than 35 lakh people comprising of over 7 lakh families. The state generates over '7500 crore income from fresh as well as dry fruits. Horticulture has emerged as an indispensible part of agriculture, offering ample opportunities for sustaining large number of agro-based industries which generate substantial employment opportunities and plays an important role in providing livelihood security to the fruit growers. Jammu & Kashmir UT is blessed with a diverse Agro climate, which is favourable for production of a variety of fruit crops like Apple, Walnut, Almond, Mango, Litchi, Cherry, Apricot, Plum, Kiwi, Olive, Citrus etc. The area which was 0.16 lakh ha under fruit cultivation in the year 1960-61 has increased to 6.59 lakh ha in year 2018-19 and the production of 0.30 lac MTs in the year 1960-61 to 97.96 lakh MTs during 2018-19. Apple and Walnut dominates the fruit production followed by Pear, Apricot, Mango, Cherry, Plum, Citrus and Almond. The Union territory of J&K has been declared as "Agri Export Zone" for Apples and Walnuts. At present, in Jammu Division, fruits occupy an area of 1.18 lakh ha with the production of 2.54 lakh MT and productivity of 2.15 MT/ha (Department of Horticulture, J&K; DES, 2019-20).

Fruit Growing Zones of Jammu Division

Sub-tropical zone

In Jammu sub-tropics, the cultivation of mango, guava, sweet orange, kinnow, lemon, lime, litchi, strawberry and low chill cultivars of peach and pear are grown. Fruits like aonla, ber, phalsa, karonda and jackfruit are grown in rainfed areas of Jammu.

Intermediate zone

This area lies in between sub-tropical zone and high hills with altitude ranging from 1000 m -1500 m above mean sea level. It includes parts of Billawar and Basohli tehsils of Kathua district as well as several localities of Rajouri, Poonch, Reasi, Udhampur, Doda, Ramban and Kishtwar districts. This zone is more suited for peaches, plums, apricots, citrus and wild pomegranate.

Temperate zone

The altitude in this zone varies between 1500 m - 2500 m and above. This zone has severe moist cold winters and experiences snow fall at higher reaches and includes area of Doda, Poonch, Ramban, Kishtwar and Rajouri and some areas of Udhampur and Kathua districts. Walnut, apple, pear, peach and plum are successfully grown in this zone.

Orchard management Practices

Orchard management is one of the most important practices for successful orcharding to influence the growth and overall development of fruit trees. The orchard management system can greatly affect earliness of bearing and yield. Management practices for orchard floor maintenance should be executed in a timely manner during the lifetime of the orchard. Orchard management practices include soil management, nutrient management, water management, canopy management and disease and pest management. Different orchard floor management systems not only suppress the growth and development of weeds but also improve soil conditions, structure and soil nutrient status as a consequence of their biomass. Orchard cultivation refers to the careful management of the orchard soil in such a way that the soil is maintained in a good condition suitable to the needs of the tree with least expenses. This involves maintenance of the physical condition of the soil, its moisture and nutrient content. Mineral nutrition coupled with different orchard soil management practices create favourable conditions for getting maximum and sustainable productivity of good quality fruits within the limits of soil, water and fertility management. Vegetation under the tree competes for nutrients and water resulting in reduced growth, yield, and small fruit. Another option is the use of organic mulches in the tree row. Examples of mulching materials include straw, wood chips, and grass residue from mowing. These mulches will suppress weed emergence, but weed removal by some means will still be necessary. Mulches can improve the water-holding capacity of some soils. Synthetic mulches made from polyethylene, polypropylene, or polyester can be placed in the tree row around the base of the trunk or as a narrow strip down the row.

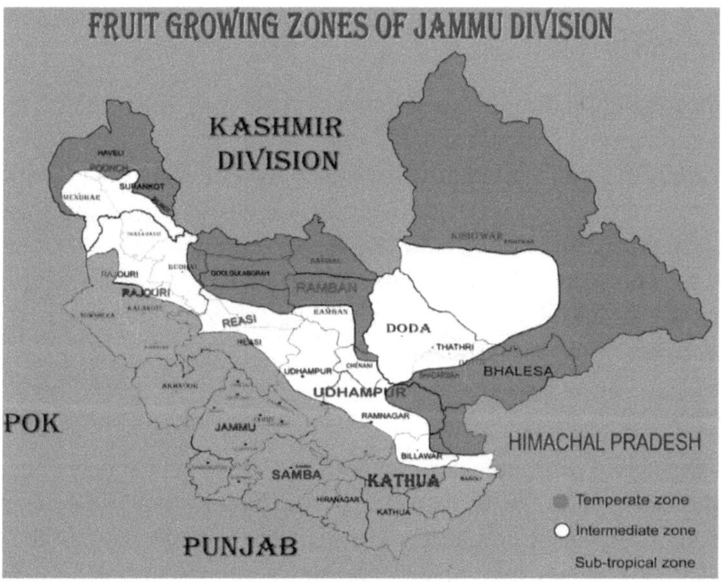

Fig. 1. Agro-climatic Zones of Jammu Division

Nursery Operations

Nursery operations are essential for providing healthy plants. It provides continuous supply of healthy and certified planting material. The important steps taken in the multiplication of quality planting material are seed extraction, seed washing, pre sowing treatment, seed sowing and after care of seedlings.

Seed extraction

The seeds should always be extracted from fully ripened fruits collected from healthy, vigorous and high yielding trees. Besides this, seed is also the main source for raising rootstock on which the plants are budded or grafted.

Washing of seeds

After extraction, the seeds should be washed properly with water and dried in shade. The seeds of sub-tropical fruits should preferably be sown immediately after shade drying. Seeds can also be stored in cool and dry places in air tight containers for certain duration.

Pre-sowing treatment

The seeds which are having very hard seed coat require scarification prior to sowing which can be done by breaking, scratching, altering or softening.

Preparation of seed beds

For raising seedlings, the bed of 2-3 m (L) x 60-100 cm (W) x 15-20 cm (H) should be prepared by mixing 2 parts loamy soil + 3 parts leaf mould + 5 parts sand. This soil media must be sterilized either by covering with white transparent plastic polyethylene sheet for 4-6 weeks during the hottest part of the year or with 2% formalin solution / basamid fungicide (300 g / cubic meter of media). This media should be mixed with 2.5 kg of base fertilizer mixture of (1.8 kg CAN or Ammonium sulphate + 2.5 kg Super phosphate + 1.0 kg Muriate of potash) per cubic meter of root media.

Seed sowing

The seeds of sub-tropical fruits should be sown immediately after extraction from the fruit, while the seeds of temperate fruits should be sown after their stratification. Since, the citrus seeds are available in winter months, the protection with alkathene tent is essential for increasing germination and the proper growth of seedlings. The seeds should be sown at a depth of 3 times the thickness of seed.

After care of seedlings

Water should be given with very fine nozzle. The frequency of irrigation in seedbed depends upon the weather and soil conditions. Hand weeding should be done to check the population of weeds. The nursery bed should be protected from summer heat by thatching with paddy straw.

Transplanting

After attaining the age of transplanting, the seedlings should be transplanted in properly leveled and well-irrigated and drained beds which have been liberally supplied with well rotten FYM/organic manures. The seedlings should be planted in twin rows at 30 cm x 15 cm distance. The distance between twin rows should be kept 50-60 cm for performing other nursery operations at a height of 20 cm from ground level.

Rootstock preparation

Select healthy seedlings for budding/grafting which have attained the diameter of thickness pencil. Remove thorns/leaves from the rootstock at budding/grafting height. The budding or grafting operations should be performed at a height of 20 cm from ground level.

Scion Preparation

Select round wood scion and it should be prepared according to the method of budding/grafting to be performed.

Planting distance

The spacing given to the fruit trees are generally governed by the various factors like climate, soil, choice of cultivars, growth habit of tree, rootstock used, nature of irrigation and pruning technique followed. Vigorous cultivars on vigorous rootstocks need more spacing than less vigorous cultivars on dwarf and semi-dwarf rootstocks.

Planting time

Evergreen plants like those of mango, citrus, guava, papaya, litchi and other fruits can be planted in two seasons *viz*., (i) Spring season (February-March) and (ii) rainy season (July-September). It is preferable to carry out planting of litchi, mango in late rainy season i.e. September or early October when weather becomes mild and more favourable and enough moisture is available in the soil. Where ample irrigation is available, spring plantation can be done. The deciduous fruits are planted once in a year i.e. during January to March depending on growing elevation when they are still dormant. Planting should be completed before emergence of new growth in spring. In snow bound areas, planting operation can be deferred till the snow melts or is cleared.

Planting of fruit trees should be avoided during hot and dry spells and preferably be done in the afternoon on cloudy and humid days rather than in bright sunshine and dry weather.

Providing windbreaks

An effective windbreak of tall trees alternating with relatively small but spreading trees must be established on the windward side of orchard, well before the orchard is planted to protect trees from hot winds of summer which damage the trees resulting in fruit splitting and fruit-drop. *Eucalyptus*, jamun, seedling mango, mulberry, jackfruit etc. are some of the trees which can make good windbreaks under sub-tropical conditions.

Digging and filling of pits

Preparation of pits and planting should always be done with the help of planting board. The guide pegs are planted at both ends. One metre deep pits of 1 metre diameter should be dug. Top 30 cm soil should be kept on one side and is used for refilling the pits as fertile soil. Bottom 70 cm soil should be kept on other side. The pits should be exposed for a few days before planting. The pits should be filled with mixture of top soil, 2-3 baskets of silt as well as well rotten farm yard manure. The refilled pits should be watered properly before planting the tree. To each pit, 0.05% chlorpyriphos can be added against white ants.

Varietal wealth

The choice of cultivars depends upon its climatic requirement, soil conditions and moisture availability. In case of temperate fruits, it is necessary to see whether the cultivars to be planted will get necessary chilling hours in winter.

Propagation

Propagation of fruit crops is done mainly by sexual and asexual methods. Sexual means through seeds and asexual means through cutting, layering, budding, grafting, micro-propagation and through specialized vegetative structures. The fruit crops which are commercially propagated through seeds are papaya and phalsa.

Nutrition

Fruit trees remove large amount of nutrients from soil than most other field crops. An orchard can be maintained in commercial bearing using adequate manuring. However, the manurial requirements of fruit trees are governed by many factors like type of fruit, cultivars, climatic conditions and cultural practices followed. Organic manures should be applied 3 months ahead of spring so that nutrients are converted into available form by the onset of spring season. The fertilizers should be placed in bands or trenches around the tree in basins. The size of basin should be increased with the increasing age and canopy of the tree.

Irrigation

For obtaining a good yield, it is necessary to have optimum soil moisture at certain critical stages of growth, flowering, fruit set and fruit development. The aim of irrigating a tree is to wet the entire rooting zone. Depending upon the availability of water, different irrigation systems, like furrow, basin, sprinkler and drip can be utilized. In rainfed areas, water-harvesting techniques during rains should be practiced.

Canopy management

Canopy management is one of the most important management practices in an orchard. It is the art of fruit growing involving manipulation of tree canopies to optimize the production of quality fruits. In fact, it forms the basis of precociousness, productivity and longevity of the fruit trees. Canopy management is an important aspect of fruit culture due to its effect on yield, quality, vigour and the prevention of diseases. Training and pruning are two important aspects of canopy management. Both affects the tree shape and facilitate exposure of leaves to sunlight. Training controls the shape, size and direction of plant growth, while pruning refers to removal of plant part like bud, shoot, root etc. to strike a balance between vegetative growth and production.

Disease and Insect-pest management

Fruit crops are prone to various diseases, insect pests and physiological disorders. So, various fungicides and insecticides are being used for their management.

Harvesting and Yield

Harvesting of the fruit crops should be done on the basis of maturity indices. Yield performance depends on the average fruit yield per tree.

2

Cultivation, Management and Economic Analysis of Mango

Mango (*Mangifera indica*), also known as king of fruits belongs to family Anacardiaceae is originated from Indo-Burma region. In India, mangoes are mainly grown in tropical and sub-tropical regions from sea level to an altitude of 1,500 m. It is the most important commercially grown fruit crop of the country. India has the richest collection of mango cultivars. In India, major mango producing states are Andhra Pradesh, Bihar, Gujarat, Karnataka, Maharashtra, Orissa, Tamil Nadu, Uttar Pradesh and West Bengal. In India, the area under mango cultivation is 2.28 million ha with annual production of 21.25 million metric tonnes. It is one of the most promising fruit crops of Jammu region, occupying an area of 13037 ha with production of 30478 metric tons (Directorate of Horticulture, Jammu, 2019)

Climate

Mango is cultivated in sub tropical and tropical conditions as it can tolerate temperature as high as 48^0 C. High humidity and cloudy weather at the time of flowering is not favourable because they affect pollination and fruit set and increase the risk of diseases. Rains during flowering are detrimental to the crop. The favourable temperature is 24^0 C to 27^0 C.

Soil

Sandy loam soils are ideal for mango cultivation. Soil pH range of 5.5 to 7.5 is desirable. Alkaline, black and saline soils are not fit for mango plantation.

Varietal Description

The popular commercial cultivars of Northern region are Dashehari, Langra, Chausa and Bombay Green, Amrapali. The other commercial cultivars like Himsagar, Langra, Fazli, Lakshman Bhog, Krishanbhog, Gulabkhas, Alphonso, Kesar, Bangalora, Swaranrekha, Mulgoa etc. are cultivated in other regions of the India.

Dashehari

Dashehari is a commercial variety of North India. The tree is medium in size. Fruit is small to medium, skin medium thick and yellow in colour. The flesh is firm, fibreless with good fruit quality. Dashehari is irregular in bearer, susceptible to hoppers and powdery mildew. It ripens in the first fortnight of July.

Alphonso

The fruits are large in size, attractive pinkish blush towards the basal end, medium bearer. Mostly suitable for the cultivation on the west coast of Maharashtra in Ratnagiri. Fruit is very tasty and rich in flavour.

Fazli

Heavy bearer with very large size fruit, matures in end of August. Quality of the fruit is poor.

Langra

Fruits are medium to large in size, flesh is firm, fibreless, lemon yellow and strongly flavoured. This is an important variety of North India with biennial bearing habits which ripens in the last fortnight of July.

Bangalora

Fruits are oblong large sized having apricot yellow colour. It is mainly grown in the southern region of India. It is regular bearer with good keeping quality.

Krishanabhog

The fruits are round in shape, medium to large size. Light yellow in colour. Ripens during the middle of July.

Chausa

Fruits are large with light yellow colour having soft and sweet pulp. Shy bearer, late maturing variety. Ripens in first fortnight of August. Highly susceptible to mango malformation.

Hybrids

Amrapali

It is a cross between Dashehari and Neelum. Amrapali is precocious, distinctly dwarf, highly regular and prolific in bearing and has good fruit quality.

Mallika

It is a cross between Neelum and Dahsehari. The tree is medium in size and regular bearer. The fruits are large, elongated, good quality and bears in cluster.

Propagation

Mango is commonly propagated by epicotyl/ stone and veneer grafting. Grafting can be done from March to April and from mid July to September.

Planting

In rainfed areas, planting is being done at the beginning of the monsoon (June-July). In areas with heavy rainfall planting should be done in the month of February-March. The pit of size 1m x 1m x 1m should be dug and filled with top soil, well mixed with 20 to 25 Kg farmyard manure. Planting distance varies with the cultivars, however 8-10 m distance is required in traditional planting. In high density plantation, the planting distance for mango is 3.0 m x 3.0 m for Amrapali and 4.5 m x 4.5 m for Dashehari.

Irrigation

Young trees are irrigated frequently, three to four days in summer and once a fortnight during winter. In case of bearing trees, irrigation at regular intervals is necessary during fruit development period.

Manuring and Fertilization

For first year, FYM @ 5 kg/tree. NPK requirement in the form of urea, DAP, MOP is 95, 50 and 35 gram/tree, respectively. The dosage need to be increased by the same amount each year. Apply full nitrogenous fertilizers alongwith half of phosphorous and half of potash after harvesting of fruit. Remaining quantities of two fertilizers should be applied during October with last irrigation.

Training and Pruning

Training of the tree for well spaced branches is essential in early years. The main branches should grow in different directions at least 30 cm apart and with good crotch angles. As mango bears terminally, so annual pruning is not done except removing over crowded, diseased and dead branches.

Weed control

Mulching with organic and inorganic mulches reduce the weed population in mango orchards, thus enhances the yield and productivity by reducing the weed population and increasing the soil moisture conditions. Black polythene mulch has been found very effective in weed control.

Harvest and Yield

The yield of the mango fruit varies due to several factors *viz*., age of the tree, variety grown, climatic conditions, soil type, type of tree (seedling or grafted one), on and off year and management practices followed. However, from a well grown mango tree (10 year onwards) the yield varies from 40 to 100 kg and may go up to 3-5 quintals per tree.

National Problems

Fruit drop

Symptoms and Causes

The natural fruit drop of mango is rather severe amounting to about 99% at various stages of growth. Lack of pollination, defective perfect flowers, self incompatibility and lack of irrigation, unfavourable climatic conditions during fruit development stage are the causes of fruit drop.

Management

Fruit drop can be reduced by applying regular irrigation during fruit development stage. Application of Naphthalene Acetic acid (NAA) @ 20 ppm (i.e. 20 mg per liter) reduces fruit drop after fruit set. Micronutrient sprays of zinc sulphate @ 0.8 per cent (i.e. 8 g per liter) and potassium sulphate @ 4% (i.e. 40 g per liter) at bloom stage helps in retention of fruits.

Alternate bearing

Symptoms and Causes

This is one of the most serious problems in the northern cultivars of mango. It means yield variation in alternate years, heavy bearing followed by less or no bearing. The causes of alternate bearing are climatic factors, age and size of shoots, C/N ratio and hormonal balances.

Management

Proper orchard management practices are to be followed. Reduce the crop load in on year by deblossoming to obtain crop in every year. Smudging and chemical regulation can help in inducing flowering and fruit set in mango. Use of paclobutrazol (10 g/tree) applied annually increases mango fruit yields. Use of regular cultivars like Amrapali and Mallika.

Mango malformation

Symptoms and Causes

The incidence of malformation varies from year to year, cultivar to cultivar and also with age of tree. Vegetative parts of the tree show formation of small shoots and short internodes at apical portion, giving appearance like witches broom. Malformed panicles become deformed, inflorescence looks like a cluster. Various causes like cultural practices, nutritional factors, hormonal imbalances, insect pests etc have been attributed this problem.

Management

Single spray of NAA @ 200 ppm (i.e. 200 mg per liter) in the first week of October and deblossming once at the bud burst stage can lower the incidence of malformation.

Black tip in mango

Symptoms and Causes

The disorder is characterized by depressed spot of yellowish tissue at distant end of fruit which increases in size becoming brown and finally black. Dashehari variety has been found to be very susceptible. The cause of black tip in mango is attributed to coal fumes and gases of brick kilns.

Management

Avoid plantation of mango orchards near brick kilns. Spraying of 0.6 per cent (i.e. 6 g per liter) borax three times *viz.*, before flowering, during flowering and after fruit setting is necessary.

Spongy tissue

The fruits have a bad odour and become unpalatable. This disorder is common in the Alphonso cultivar. The symptoms are only visible when the fruit is cut. Reports prevail that disorder is caused by convective heat rising from the soil which directly hits the exposed fruits during their development and ripening stage.

Management

Use of cover crops and mulches are beneficial for reducing the incidence of this disorder.

Diseases

Powdery mildew

Symptoms and Causes

Powdery mildew (*Oidium mangiferae*) whitish or greyish powdery mass appears on young foliage and inflorescence. The affected panicles bear very little or no fruit due to pre-mature withering of inflorescence. Fruit drop may also take place.

Management

Spray dinocap (50 g/100 Lwater), carbendazim (100 g/100 L water), or wettable sulphur (3 g/1 L water) when disease appears. Give 2-3 subsequent sprays at an interval of 10-15 days as per disease severity.

Anthracnose

Symptoms

Anthracnose (*Colletotrichum gloeosporioides*) disease appears on young leaves, shoots inflorescence and fruits. Round or irregular brown spots appear on leaves and young shoots. Spots on leaves enlarge with cracked centre. Spots on shoots enlarge longitudinally giving a blighted appearance. In severe attack, the shoots may die. Minute black dots appear on inflorescence and fruit skin. Affected fruit becomes sunken.

Management

Prune off the diseased/dead shoots and apply Bordeaux paste or copper oxychloride paste to cut ends during December-January. Spray the trees with mancozeb (200 g/100 L water), copper oxychloride (300 g/100 L water), carbendazim (100 g/100 L water) or captan (300 g/100 L water). Give one spray each before and after flowering and 2-3 subsequent sprays at fortnightly intervals as the monsoon sets in.

Bacterial Canker

Symptoms

Bacterial Canker (*Xanthomonas campestris* pv. Mangiferae) show minute, dark green, water soaked areas appear on leaves and fruits. Later, these spots become raised. The disease spread is rapid during rains and becomes severe.

Management

Avoid overcrowding of the trees and water logging. Mango orchards should be properly maintained by pruning the intermingling branches for free circulation of air and sunlight to minimize the congenial condition for mango hopper breeding. Orchards may be sprayed thrice as given below:

1st spray: End of February use Dichlorvos 100 EC 500 ml or Carbaryl 50 WP, 1000 g in 500 L water.

2nd spray: End of March after fruit set, use Dimethoate 30 EC 750 ml of Methyl demeton 25 EC 800 ml or Monocrotophos 36 EC 600 ml in 500 L water.

3rd spray: May be given in April or four weeks after 2nd spray.

Insect-Pests

Mango hopper

Symptoms

This pest is active from February to May. Damage is caused both by nymphs and adults by sucking sap from tender leaves, shoot in inflorescences. As a result, fruit setting is affected and quality is reduced. Nymphs while feeding secrete honey dew which falls on leaves and encourages the growth of sooty mould fungus.

Management

Avoid overcrowding of the trees and water logging. Mango orchards should be properly maintained by pruning the intermingling branches for free circulation of air and sunlight to minimize the congenial condition for mango hopper breeding. Orchards may be sprayed thrice as given below:
1st spray: End of February use Dichlorvos 100 EC 500 ml or Carbaryl 50 WP, 1000 g in 500 L water.

2nd spray: End of March after fruit set, use Dimethoate 30 EC 750 ml of Methyl demeton 25 EC 800 ml or Monocrotophos 36 EC 600 ml in 500 L water.
3rd spray: May be given in April or four weeks after 2nd spray.

Mango shoot borer

Symptoms

Larvae of this pest after hatching from the egg bores into the young shoot from top to down-wards and makes small gallery and feeds on the soft tissues.The attacked shoots show dropping and leaves dry.

Management

Remove all the dried shoots and destroy them. Spray the new growth once or twice in August- September with or Methylparathion 50 EC, 600 ml in 500 L water.

Mango mealy bug

Symptoms

Mango mealy bug (*Drosicha mangiferae*) damage is rendered by nymphs and adult females by sucking sap from young shoots and inflorescence, panicles as a result fruit setting is reduced and fruits remain undersize and some even fall to the ground.

Management

Prune the affected parts and burn the same in the initial stage of infestation. Soil in the mango orchard should be turned over in June-July with a furrow turning plough to expose the egg masses and their destruction by summer heat. Treat the basins of the trees and area around the mango tree in January-February with 1.5 per cent lindane dust @ 25 to 30 kg/ha. Wrap the tree trunk with alkathene bands 15-20 cm wide (400 gauge) about ½ m above ground and tightly fastened to the trunk before eggs start hatching in February to prevent the ascending of nymphs. Apply mud plaster at the lower side to prevent the nymphs from ascending through the band. The nymphs which have ascended and settled on the growing points can be checked by sprayingcarbaryl 50 WP, 1000 g in 500 L water.

Economic Analysis

The information with respect to yearly costs involved in the establishment of mango orchards cv. Dashehari with spacing of 10 m × 10 m are as:

Cost of establishment

During the first year of the plantation, orchardists had incurred expenditure on the preparation of land, planting material, digging, filling of pits and planting, irrigation, plant protection, farm yard manure, fertilizers, intercultural operations etc. The operation wise per hectare average costs for the first year is presented in Table 6. The per hectare total costs in first year for the establishment of mango orchards were found to be ₹ 28859.82 per hectare. Variable costs per hectare were found to be ₹ 22699.82 with ₹ 7500.00 incurred on digging, filling and planting, ₹ 3000.00 on preparation of land, ₹ 4200.00 on planting material, ₹ 1850.00 on irrigation, ₹ 600.00 on farm yard manure, ₹ 617.70 on plant protection chemicals, and ₹ 1800.00 on intercultural operations whereas fixed costs were equal to ₹ 6160.00 per hectare with ₹ 2250.00 incurred on implements, ₹ 3250 as earned value of rented land (EVRL). The per hectare interest on working and fixed capital was worked out to ₹ 2432.12 and ₹ 660.00, respectively.

Operational costs

The per hectare operational costs consists of yearly expenses on the maintenance of bearing orchard. The item wise and concept wise operational costs for mango orchard are presented in Table 7. The per hectare variable costs and fixed were found to be ₹ 27022.18 and ₹ 5320.00, respectively with a total of ₹ 32342.18 per hectare. The table further revealed that among the working costs, the highest expenditures of ₹ 12000.00 per hectare was incurred on farm yard manure and lowest of ₹ 1746.95 was found to be incurred on fertilizers. The fixed costs included ₹ 1500.00 on depreciation on machinery and farm inventory, ₹ 3250.00 as earned value of rented land and ₹ 570.00 as interest on fixed capital

Year wise cost and return analysis

The per hectare per year cost and returns from mango orchards worked out for different age groups are presented in Table 8. The returns were worked out upto 7th year of the age of an orchard, between 8th– 14th year and above 15th year of orchard. These groups were made on the fact that the returns upto 7th year were low as compared to other groups, whereas from 8th– 14th year the returns were increasing and from 15th year onwards the returns almost remained constant.

The returns started from 5th year of the establishment of the orchard with production of 15 quintals per hectare which increases to 80 quintals per hectare from 15th year and onwards. As far as the net returns are concerned, it was found to be ₹ 22467.73 per hectare in 5th year which increases to ₹ 207657.82 per hectare during 15th year of the establishment of the orchard and onwards. The benefit cost ratio of 1:2 was found in 5th year which increases to 1:7.42 during 15th year of the establishment of the orchard which proves the viability of the orchard.

Table 6 Operation wise first year establishment costs of mango orchards cv. Dashehari and spacing = 10 m x 10 m (₹/ha) No. of Plants per hectare = 100

S.No.	Item	Cost (in ₹)
1	Preparation of land	3000.00
2	Digging, filling and planting	7500.00
3	Planting material/ Plantation cost	4200.00
4	Irrigation	1850.00
5	Farm Yard Manure	600.00
6	Fertilizers (Urea, DAP, MoP)	617.70
7	Plant Protection Chemicals	700.00
8	Intercultural operations	1800.00
9	Interest on working capital @12 per cent per annum	2432.12
A	Total variable cost (1-9)	22699.82
10	Implements	2250.00
11	Earned value of rented land	3250.00
12	Interest on fixed capital @ 12 per cent per annum	660.00
B	Total fixed cost (10-12)	6160.00
13	Total establishment cost (A+B)	28859.82

Table 7 Item wise and concept wise operational costs of mango orchards cv. Dashehari and spacing = 10 m x 10 m (₹/ha)

S.No.	Item	Cost (in ₹)
1	Labour	12000.00
2	Intercultural operations and hoeing, weeding	3000.00
3	Irrigation	1500.00
4	Training/ pruning	1800.00
5	Farm Yard Manure	12000.00
6	Fertilizers (Urea, DAP, MoP)	1746.95
7	Plant protection	2200.00

8	Interest on working capital	2895.23
A	Total variable cost (1-8)	27022.18
9	Depreciation	1500.00
10	EVRL*	3250.00
11	Interest on fixed capital	570.00
B	Total fixed cost (9-11)	5320.00
12	Total operational/ maintenance cost (A+B)	32342.18

Table 8 Year wise cost and return analysis of mango orchards cv. Dashehari and spacing = 10 m x 10 m (₹/ha)

S.No.	Particulars	Years								
		1	**2**	**3**	**4**	**5**	**6**	**7**	**Average 8th to 14th year**	**15th year and onwards**
1.	Total cost	28859.82	18441.25	19804.62	21168.45	22532.27	23896.10	25259.64	27988.54	32342.18
2.	Production (qtls)	0.00	0.00	0.00	0.00	15	21	27	52	80
3.	Gross returns (in ₹)	0.00	0.00	0.00	0.00	45000.00	63000.00	81000.00	156000.00	240000.00
4.	Net returns (in ₹)	-28859.82	-18441.25	-19804.62	-21168.45	22467.73	39103.90	55740.36	128011.46	207657.82
5.	Benefit cost ratio	0.00	0.00	0.00	0.00	1:2.00	1:2.64	1:3.21	1:5.57	1:7.42

Economic analysis of high density mango orchards cv. Dashehari

The information with respect to yearly costs involved in the establishment of high density mango orchards cv. Dashehari with spacing of 4.5 m × 4.5 m are as:

Cost of establishment

During the first year of the plantation, orchardists had incurred expenditure on the preparation of land, planting material, digging, filling of pits and planting, irrigation, plant protection, farm yard manure, fertilizers, intercultural operations etc. The operation wise per hectare average costs for the first year is presented in Table 9. The per hectare total costs in first year for the establishment of high density mango orchards were found to be ₹ 77883.12.Variable costs per hectare were found to be ₹ 70323.12 with ₹ 29640.00 incurred on digging, filling and planting, ₹ 3000.00 on preparation of land, ₹ 20748.00 on planting material, ₹ 1850.00 on irrigation, ₹ 2964.00 on farm yard manure, ₹ 1200.00 on plant protection chemicals, and ₹ 2500.00 on intercultural operations whereas fixed costs were equal to ₹ 7560.00 per hectare with ₹ 3500.00 incurred on implements and ₹ 3250 as earned value of rented land (EVRL). The per hectare interest on working and fixed capital was worked out to ₹ 7534.62 and ₹ 660.00, respectively.

Operational costs

The per hectare operational costs consists of yearly expenses on the maintenance of bearing orchard. The item wise and concept wise operational costs for high density mango orchard are presented in Table 10. The per hectare variable costs and fixed were found to be ₹ 97899.12 and ₹ 5656.00, respectively with a total of ₹ 103555.12 per hectare. The table further revealed that among the working costs, the highest expenditures of ₹ 59280.00 per hectare was incurred on farm yard manure and lowest of ₹ 1800.00 was found to be incurred on irrigation charges. The fixed costs included ₹ 1800.00 on depreciation on machinery and farm inventory, ₹ 3250.00 as earned value of rented land and ₹ 606.00 as interest on fixed capital

Year wise cost and return analysis

The per hectare per year cost and returns from high density mango orchards worked out for different age groups are presented in Table 11.The returns were worked out upto 7th year of the age of an orchard, between 8th– 14th year and above 15th year of orchard. These groups were made on the fact that the returns upto 7th year were low as compared to other groups, whereas from 8th–14th year the returns were increasing and from 15th year onwards the returns almost remained constant.

The returns started from 4th year of the establishment of the orchard with production of 98.80 quintals per hectare which increases to 592.80 quintals per hectare from 15th year and onwards. As far as the net returns are concerned, it was found to be '230919.45 per hectare in 4th year which increases to '1674844.88 per hectare during 15th year of the establishment of the orchard and onwards. The benefit cost ratio of 1:4.53 was found in 4th year which increases to 1:16.17 during 15th year of the establishment of the orchard which proves the viability of the orchard.

Table 9 Operation wise first year establishment costs forhigh density mango orchards cv. Dashehari and spacing= 4.5 m x 4.5 m (₹/ha) No. of Plants per hectare = 494

S.No.	Item	Cost (in ₹)
1	Preparation of land	3000.00
2	Digging, filling and planting	29640.00
3	Planting material/ Plantation cost	20748.00
4	Irrigation	1850.00
5	Farm Yard Manure	2964.00
6	Fertilizers (Urea, DAP, MoP)	886.50
7	Plant Protection Chemicals	1200.00
8	Intercultural operations	2500.00
9	Interest on working capital @12 per cent per annum	7534.62
A	Total variable cost (1-9)	70323.12
10	Implements	3500.00
11	Earned value of rented land	3250.00
12	Interest on fixed capital @ 12 per cent per annum	810.00
B	Total fixed cost (10-12)	7560.00
13	Total establishment cost (A+B)	77883.12

Table 10 Item wise and concept wise operational costs for high density mango orchards cv. Dashehari and spacing= 4.5 m x 4.5 m (₹/ha)

S.No.	Item	Cost (in ₹)
1	Labour	3700.00
2	Intercultural operations and hoeing, weeding	7000.00
3	Irrigation	1800.00
4	Training/ pruning	2500.00
5	Farm Yard Manure	59280.00
6	Fertilizers (Urea, DAP, MoP)	8629.93
7	Plant protection	4500.00
8	Interest on working capital	10489.19
A	Total variable cost (1-8)	97899.12
9	Depreciation	1800.00
10	EVRL*	3250.00
11	Interest on fixed capital	606.00
B	Total fixed cost (9-11)	5656.00
12	Total operational/ maintenance cost (A+B)	103555.12

Table 11 Year wise cost and return analysis for high density mango orchards cv. Dashehari and spacing= 4.5 m x 4.5 m (₹/ha)

S.No.	Particulars	Years								
		1	**2**	**3**	**4**	**5**	**6**	**7**	**Average 8th-14th year**	**15th year and onwards**
1.	Total cost	77883.12	58565.25	62745.42	65480.55	69540.72	73965.50	80925.44	87889.46	103555.12
2.	Production (qtls)	0.00	0.00	0.00	98.80	133.38	172.90	212.42	397.32	592.80
3.	Gross returns (in ₹)	0.00	0.00	0.00	296400.00	400140.00	518700.00	637260.00	1191951.30	1778400.00
4.	Net returns (in ₹)	-77883.12	-58565.25	-62745.42	230919.45	330599.28	444734.50	556334.56	1104061.84	1674844.88
5.	Benefit cost ratio	0.00	0.00	0.00	1:4.53	1:5.75	1:7.01	1:7.87	1:13.56	1:16.17

3

Cultivation, Management and Economic Analysis of Citrus Fruits

Citrus (*Citrus* spp.) is one of the most important fruits grown in sub-tropical and foot hill regions of Jammu province (Kathua, Hiranagar, Samba, Jammu, Akhnoor, parts of Basohli, Nowshera, Ramnagar, Ramban and Udhampur) upto an elevation of 800 m amsl. The most important commercial citrus fruits in Jammu region are kinnow mandarin, sweet orange, grapefruit, lemon and lime. The area under citrus fruits in Jammu division of state is 14,542 ha with an annual production of 34,191 metric tones.

Climate and soil

Citrus plants being sub-tropical cannot with stand extended cold period. Temperature below 3°C and above 40°C are injurious to the citrus. Hot winds and excessive heat during flowering and fruit set cause fruit drop and sun burning. These are the light loving crops.

Citrus fruits thrive well in deep, loose, well aerated soils with the pH of 5.5 to 6.2 devoid of any hard pan layers of calcium carbonate in the rooting zone up to 150 cm. High water table and poor drainage system are quite unsuitable for citrus fruits which cause drying of roots and nutritional imbalance, respectively.

Cultivars

Fruit	Cultivar	Characteristics
Mandarin (*C.reticulata*)	**Kinnow**	Fruit medium in size, globose to oblate, golden orange colour when fully ripe, strong aroma, moderately acidic, tight skinned and ripens in December.
	Local Santra	Fruit medium to small size, oblate to sub-globose, loose skin, fair flavour, juicy, slightly acidic and ripens in January.
Sweet orange (*C. sinensis*)	**Mosambi**	Fruit medium in size, sub-globose shape, rough skin with longitudinal furrows, apex marked with a typical circular ring, flesh pale-yellow or whitish with low acid content, and ripens in November.
	Pineapple	Fruit medium to large, round to slightly oblate, smooth skin with deep orange colour, juicy with excellent flavour and ripens in December.
	Jaffa Malta	Medium to large fruit, round to oblate, golden yellow to orange red skin, juicy with rich flavour and ripens in December.
	Blood Red	Medium to large fruit, roundish to slightly oblong with thin skin, orange yellow to deep orange colour, fully red flesh when ripens, rich in flavour and ripens in January.
	Valencia	Medium to large fruit, slightly oval, golden yellow in colour, juicy sub-acidic in taste and ripens in February.
Grapefruit (*C.paradisi*)	**Marsh Seedless**	Fruit medium to large, roundish oblate, light yellow colour with smooth skin and ripens in January.
	Duncan	Fruit large, oblate, light yellow or orange in colour, fruits quite acidic and sweet but bitter and ripens in January.
	Foster	Medium to large fruit, oblate in shape, pale yellow with pink flesh, well blended with acidity and sweetness along with well marked bitterness, and ripens in November-December.
Lemon (*C.limon*)	**Eureka**	Fruits medium in size, oblong with a nippled apex, yellow skin, juicy, acidic with excellent flavour and ripens in August-September.

	Baramasi	Bear fruits in 3-4 flushes throughout the year, fruits round and yellow with a round apex, smooth and thin skin, juicy and almost seedless.
	Galgal	Large oval, yellow at maturity and suitable for making pickles, squash and candy.
Acid lime (Nimboo) (*C.aurantifolia*)	**Kaghzi**	Fruits small, round and thin skinned, greenish white, juicy and highly acidic.
Sweet lime (*C.limettoides*)	**Local**	Fruits medium to large, pale yellow with smooth skin and typical aroma, juicy and ripens in September.

Propagation and Rootstock

Most of the citrus fruits are generally propagated through T-budding. Jatti Khatti (*Citrus jambhiri*) is the common rootstock for most of the citrus fruits but Jatti Khatti and Rangpur lime for Kinnow, Rangpur lime and Cleopatra mandarin for Blood Red, Jatti Khatti and Cleopatra mandarin for Jaffa are promising. The fresh seeds of rootstocks are sown in beds at 2 cm x 10 cm distance and 1.0 cm deep which germinate 3 weeks after sowing. After attaining the 15 cm height, the seedlings should be transplanted in nursery where budding operations are to be performed. Citrus plants can be budded either during March-April or August-September. Lime and lemons are propagated through air layering in rainy season.

Planting

a) Planting Time: July-August

February-March

b) Planting distance:

Lemon and lime	5-6 m
Grapefruit	6-7 m
Mandarin and sweet orange	6-9 m

Citrus plants are planted in pits of 50 cm x 50 cm x 50 cm or 75 cm x 75 cm x 75 cm sizes and irrigated thereafter. The young plants should be protected from cold weather during winter season.

Training and Pruning

Thinning out of branches in young nursery plants should be done at least once a month during the year following planting and once in two to three months during the first three years just for avoiding very low headed trees. Training should be done to develop mechanically strong trees with well spaced scaffold limbs at the early years before fruiting. In bearing trees, ground touching, diseased or dead branches should be removed after harvesting of fruits.

Manuring and Fertilization

Following schedule should be followed for fertilizing citrus trees.

Age of tree (Year)	**FYM (Kg/tree)**	**Fertilizer requirement (g/tree)**		
		Urea	**Diammonium Phosphate**	**Muriate of potash**
1	5	160	35	25
2	10	320	70	50
3	15	480	105	75
4	20	640	140	100
5	25	800	170	120
6	30	960	205	150
7	35	1120	240	175
8	40	1280	275	200
9	45	1445	310	225
10 Year and above	50	1600	345	250

The full amount of FYM and phosphatic fertilizer should be applied in January. Whole potash and ½ nitrogenous fertilizers should be applied 15 days before flowering, whereas remaining ½ of the nitrogenous fertilizer should be given after fruit set.

Foliar spray of nutrient mixture (400 g $CuSO_4$ + 200 g $FeSO_4$ + 200 g borax + 1.04 kg lime + 100 L water) on new emerging growth flushes is beneficial for improving yield and quality of fruits.

Irrigation

Citrus trees should be irrigated at 10 days and 30 days interval during summer and winter, respectively. Irrigation water should not come in direct contact with tree trunk. The young plants (up to 3-4 years) should be irrigated at weekly intervals.

Intercropping

Intercropping in citrus orchards up to age of 3-4 years is desirable, for which shallow rooted, short duration, less exhaustive having good canopy cover and helpful in adding nutrients through nitrogen fixing crops (pea, cowpea, gram, beans etc.) are considered safe and profitable. In addition to that, some other crops like tomato, okra etc. can also be grown as the intercrops in non-bearing citrus orchards.

Flowering and fruit thinning

In mandarin, sweet orange and grapefruits, flowering takes place in spring, while lemon and limes flower almost round the year. Generally, citrus fruits are self-pollinated, but cross pollination is not uncommon. Honey bees are known to act as the pollinating agent in citrus. Fruit thinning is essential in overbearing trees of citrus fruits (like kinnow) after fruit set in May in order to maintain fruit size and tree health and to ensure regular cropping.

Harvesting and yield

Citrus fruits should be harvested when they attain full size, ground colour and recommended TSS/acid ratio. Citrus fruits are harvested generally between December to February. The fruits should be clipped from the tree retaining a non-protruding short fruit stalk but not by pulling from the branches. Average yield in terms of number of fruits for different citrus fruits is given as under:

Sweet orange	-	400-500 fruits/tree/year
Mandarin	-	1000-1500 fruits/tree/year
Kinnow	-	300-500 fruits/tree/year
Grape fruit	-	150-300 fruits/tree/year
Lime	-	800-1000 fruits/tree/year
Lemon	-	500 fruits /tree/year

Special Problems

Diseases and their Management

Symptoms	Management
Citrus Canker: (*Xanthomonas campestris* pv. *citri*) Dark brown, rough, raised, pimple-like corky spots appear on leaves, twigs, branches, petioles, fruit stalks, fruits and thorns. On leaves, spots first appear on lower surface and then on upper surface, spots are surrounded by a yellow halo. Lesions on twigs are elongated. On fruits, the lesions are confined to rind only.	Prune off the affected twigs before monsoon and destroy by burning. Spray the tree and nursery plants with copper oxychloride (300 g/100 L water) + streptomycin sulphate (50g/ 100 L water) or streptocycline (100g/100L water) at 15 days interval, give 5-7 sprays.
Gummosis or Foot Rot: (*Phytophthora* spp.) Bark on the tree trunk at the soil level turns brown, develops longitudinal cracks and starts gumming. The affected trees become weak, exhibit sickly appearance, usually blossom heavily and die before the fruit matures due to girdling effect.	i) Select plants from disease free nursery. ii) While planting, keep bud union above the soil. iii) Scrap the disease portion along with some healthy part with sharp knife. Disinfect the wound with mercuric chloride (1g/L water) and apply Bordeaux paste (2:3:30) or copper oxychloride paste. iv) Maintain good water drainage in the orchard. Avoid excessive irrigation, stagnation of water and injury to root or base of the tree trunk.
Melanose: (*Phomopsis citri*) Minute, brown, circular spots with yellow margins appear on young leaves which late become raised and turn dark brown. Similar lesions also appear on twigs. The affected fruits show small and tough pustules confined to rind only.	Prune off and burn the diseased twigs. Spray with copper oxychloride (300g/100L water) or thiophenate methyl (100 g/100L water) twice during March to April at 15-20 day interval. Repeat 2-3 sprays at fortnightly interval starting immediately after first monsoon shower.
Wither tip/Anthracnose: (*Colletotrichum gloeosporioides*) Wither tip is characterized by shedding of leaves and die back of twigs. In severe infection, branches	i) Maintain tree vigour by proper irrigation, manuring and cultural practices. ii) Prune off dead twigs during February and destroy by burning. Protect the cut ends with Bordeaux paste or copper oxychloride

show die back symptoms and the tree dies in a few years. On leaves, necrotic spots are found in concentric rings.	paste. iii) Spray with copper oxychloride (300g/L water) or carbendazim (100g/100L water) during March, July and September. Repeat sprays at 15 days interval as per disease severity.
Citrus decline: (*A syndrome involving infection by fungi, nematodes, viruses, phytoplasma* and *nutritional disorders*). The trees show decline in health and fruit yield, growth is stunted and the trees may ultimately die.	i) Adopt cultural practices like hoeing, proper fertilization and irrigation. ii) Control aphids and citrus psylla by spraying insecticides like dimethoate, phosphomidon, monocrotophos or aldicarb to check the disease. iii) Spray carbendazim (100g/100L water), copper oxychloride (300g/100L water), mancozeb (250g/100L water) or thiophanate methyl (100ml/100L water) for controlling fungal diseases. iv) Spray zinc sulphate + lime (600 g + 300g/100L water) at tender foliage stage. v) Use disease free bud-wood planting material for new plantation. vi) Adopt integrated control measures with budwood certification, use of resistant rootstocks and fungicides/insecticides where the problem is acute.
Damping off and Wilt in Nurseries: (*Pythium* sp., *Phytophthora* sp. *Fusarium* sp., *Rhizoctonia* sp.) Water-soaked spots appear on the stem at collar region. Young seedlings fall and die immediately. In old seedlings, spots increase and ultimately the seedlings die.	i) Select well-drained soil for raising seedlings and use raised beds. ii) Avoid over-crowding of seedlings, improper irrigation and infested/sick soil. iii) Use well rotten FYM compost. iv) Treat the seed bed, before sowing with 40% formaldehyde (40 ml/L water with 5L solution for sq. metre area) and cover with polythene sheet. Remove polythene sheet after 48h and plant after two weeks. v) Treat seed with carbendazim (1g/kg seed) or thiram (3g/kg seed) before sowing. vi) Burning of trash (10-12 cm thick) over the well worked soil and rotation of nursery bed help in avoiding the disease. vii) Drench the standing nursery with thiram (200g/100L water) or carbendazim (100g/100L water).

Insect Pest Management

Symptoms	Management
Citrus Leaf Miner:(*Phyllocinistis citrella*) Young larvae mine into the epidermal layer of young tender leaves by making silvery white zigzag galleries and suck sap. Infested leaves curl and their photosynthetic activity is reduced. Attacked trees bear poor quality fruit.	Prune the heavily affected parts during winter and burn the same. The peak period of these pests is March to September when maximum damage is rendered, therefore, a spray schedule comprising of 4 to 5 sprays will control these pests. 1st Spray during February-March before flowering use or dichlorvos 100 EC 700 ml or methyl parathion 50 EC 600 ml or dimethoate 30 EC 650 ml or methyl demeton 25 EC 650 ml in 500 L of water. 2nd Spray end of March or four weeks after 1stspray (after fruit set) dimethoate 30 EC 650 ml or monocrotophos 36 EC 600 ml. In 500 L of water. 3rd Spray 3 weeks after 2nd spray in end of April or May use same insecticides as recommended for 2nd spray. 4th Spray in July-August use same insecticides as recommended for 2nd spray. 5th Spray in August-September use anyone of the insecticides mentioned for 1st spray.
Citrus Psylla: (*Diaphorina citri*) Small wedge shaped grey insects seen sitting in a typical symmetry on the underside of the leaves and suck the sap, as a result, the leaves curl and distorted, the quality of fruits is lowered.	Follow same spray schedule as in citrus leaf miner.
Lemon Butterfly: (*Papilio demolus*) The caterpillars of this butterfly feed on the eaves from margin inside and defoliate the branches. Both nursery and grown up plants are attacked and their vitality is reduced.	Hand picking and destruction of various stages of the pest. Follow same spray schedule as in citrus leaf miner.
Bark Eating Caterpillars: Write the material as in mango.	Write all control measures point wise on in mango. Paste the material from mango here

Economic Analysis

Kinnnow

Kinnow being a perennial fruit crop starts bearing five years after plantation. The information with respect to yearly costs involved in the establishment of kinnow orchards is necessary in order to allocate the non recurring cost to the yearly cost of production.

Cost of establishment

Like orange, high investment is involved for the establishment of kinnow orchard during its first year with low cost involvement during the subsequent years till it starts bearing. The operation wise first year establishment costs of kinnow is presented in Table 12. In case of marginal orchards, the per hectare costs incurred were ₹ 2764.05 on preparation of land, ₹ 4075.00 on digging, filling and planting, ₹ 854.28 on planting material, ₹ 239.70 on irrigation, ₹ 95.00 on training/ pruning, ₹ 635.75 on (manures + fertilizers), ₹ 313.40 as depreciation on machinery and farm inventory and ₹ 2734.23 on earned value of rented land (EVRL). The interest on working and fixed capital worked out to be ₹ 1079.13 and ₹ 365.73, respectively, hence the total costs amounted to ₹ 13156.25. In case of small orchards, ₹ 4002.98 were spent on digging, filling and planting, ₹ 2885.45 on preparation of land, ₹ 949.23 on planting material, ₹ 210.95 on irrigation, ₹ 95.00 on training/ pruning, ₹ 28.50 on plant protection and ₹ 647.83 on (manures + fertilizers). The depreciation on machinery and farm inventory was ₹ 316.45 and EVRL worked out to be ₹ 2822.50 with interest on working and fixed capital to be ₹ 1142.00 and ₹ 376.63, respectively. The costs incurred by medium orchardists were ₹ 3813.68 on digging, filling and planting, ₹ 2526.63 on preparation of land, ₹ 895.08 on planting material, ₹ 262.88 on irrigation, ₹ 100.00 on training/ pruning, ₹ 116.48 on plant protection and ₹ 1082.20 on (manures + fertilizers) with depreciation of ₹ 413.15 and EVRL of ₹ 3077.25. The interest on working and fixed capital worked out to be ₹ 1103.63 and ₹ 418.85, respectively. The overall average first year total establishment costs per hectare were ₹ 13245.80 out of which ₹ 4051.73 were incurred on digging, filling and planting, ₹ 2784.43 on preparation of land, ₹ 876.50 on planting material, ₹ 233.95 on irrigation, ₹ 95.15 on training/ pruning, ₹ 72.49 on plant protection and ₹ 650.83 on (manures + fertilizers), ₹ 316.85 as depreciation and EVRL of ₹ 2763.38. The per hectare interest on working and fixed capital worked out to be ₹ 1093.78 and ₹ 369.63, respectively.

The year wise establishment costs per hectare of kinnow orchards are given in Table 13. The total establishment cost per hectare incurred on kinnow orchards were ₹ 31370.33 in case of marginal orchards, ₹ 32727.25 in case of small orchards and ₹ 34856.93 in case of medium orchards with an overall average of ₹ 31768.73. The data in the table further indicated that the first year establishment costs for marginal, small and medium orchards were ₹ 13156.25, ₹ 13477.50 and ₹ 13809.80, respectively, ₹ 5912.73, ₹ 6392.03 and ₹ 6875.30, respectively for the second year, ₹ 6113.28, ₹ 6424.65 and ₹ 7120.43, respectively for the third year whereas for the fourth year they were ₹ 6188.08, ₹ 6433.08 and ₹ 7051.40, respectively. The average, first, second, third and fourth year per hectare establishment costs for all the size of holdings taken together were ₹ 13245.80, ₹ 6045.98, ₹ 6210.45 and ₹ 6266.50, respectively.

Operational Costs

The item wise and concept wise operational costs are presented in Table 14. The per hectare cost of cultivation towards cost A, cost B and cost C, respectively worked out to be ₹ 2627.03, ₹ 6642.15 and ₹ 9767.15 in marginal orchards, ₹ 3575.83, ₹ 7752.30 and ₹ 10469.80 in small orchards and ₹ 3270.40, ₹ 7729.48 and ₹ 10719.48 in medium orchards, whereas per hectare overall average cost worked out to be ₹ 2855.73, ₹ 6919.05 and ₹ 9949.75, respectively. In case of marginal orchards the per hectare costs incurred were ₹ 1777.13 on hired human labour, ₹ 232.88 on training/ pruning, ₹ 294.05 on (manures + fertilizers), ₹ 41.50 on plant protection, ₹ 281.48 as interest on working capital, ₹ 308.75 as depreciation, ₹ 3276.18 as EVRL, ₹ 430.20 as interest on fixed capital and ₹ 3125.00 on family human labour. In case of small orchards, the per hectare costs incurred were ₹ 2009.25 on hired human labour, ₹ 232.88 on training/ pruning, ₹ 736.85 on (manures + fertilizers), ₹ 97.50 on plant protection, ₹ 383.13 as interest on working capital, ₹ 355.63 as depreciation, ₹ 3373.38 as EVRL, ₹ 447.48 as interest on fixed capital and ₹ 2717.50 on family human labour. Similarly, in medium orchards the per hectare costs incurred were ₹ 1851.63 on hired human labour, ₹ 491.18 on training/ pruning, ₹ 487.83 on (manures + fertilizers), ₹ 89.38 on plant protection, ₹ 350.40 as interest on working capital, ₹ 413.03 on depreciation, ₹ 3568.30 as EVRL, ₹ 477.75 as interest on fixed capital and ₹ 2990.00 on family human labour. The data further indicated that on an average the per hectare expenditures involved were ₹ 1830.78 on hired human labour, ₹ 3030.70 on family human labour, ₹ 397.83 on (manures+fertilizers), ₹ 3305.90 as earned value of rented land, ₹ 55.28 as plant protection, ₹ 265.88 as training/ pruning, ₹ 305.98 as interest on working capital, ₹ 322.08 as depreciation and ₹ 435.35 as interest on fixed capital.

Table 12 Operation wise first year establishment costs under different size groups of kinnow orchards (₹/hectare)

Item	Marginal	Small	Medium	Overall
Preparation of land	2764.05	2885.45	2526.63	2784.43
Digging, filling and planting	4075.00	4002.98	3813.68	4051.73
Planting material	854.28	949.23	895.08	876.50
Irrigation	239.70	210.95	262.88	233.95
Training/ Pruning	95.00	95.00	100.00	95.15
Manures + Fertilizers	635.75	647.83	1082.20	650.83
Plant protection	0.00	28.50	116.48	9.58
Interest on working capital	1079.13	1142.00	1103.63	1093.78
Land revenue	0.00	0.00	0.00	0.00
Depreciation	313.40	316.45	413.15	316.85
Earned value of rented land (EVRL)	2734.23	2822.50	3077.25	2763.38
Interest on fixed capital	365.73	376.63	418.85	369.63
Total	13156.25	13477.50	13809.80	13245.80

Table 13 Year wise establishment costs under different size groups of Kinnow orchards (₹/hectare)

Year	Marginal	Small	Medium	Overall
I	13156.25	13477.50	13809.80	13245.80
II	5912.73	6392.03	6875.30	6045.98
III	6113.28	6424.65	7120.43	6210.45
IV	6188.08	6433.08	7051.40	6266.50
Total	31370.33	32727.25	34856.93	31768.73

Returns

The per year per hectare returns from kinnow orchards for different age groups (upto 10 years, 11th to 15th year and above 15th year) are presented in Table 15. The returns upto10th year of age were ₹ 13815.75 in marginal orchards, ₹ 14656.63 in small orchards and ₹ 15588.88 in medium orchards. Similarly, it was ₹ 16963.30, ₹ 17385.88 and ₹ 18142.18 for marginal, small and medium orchards, respectively from 11th to 15th year while as from 15th year onwards the returns per hectare per year were ₹ 17385.80 in marginal orchards, ₹ 17813.38 in small orchards and ₹ 19913.75 in medium orchards. The average returns per hectare per year from marginal, small and medium orchards were ₹ 14051.88, ₹ 17089.95 and ₹ 17551.05, respectively with an overall average of ₹ 16580.18 for the whole life span of the kinnow orchard.

Economic Viability

The economic viability of kinnow orchards is presented in Table 16. The net present value was worked out to be ₹ 18668.83 in marginal orchards, ₹ 23469.48 in small orchards and ₹ 29123.08 in medium orchards. The internal rate of return was 14.75 per cent in marginal orchards, 15.50 per cent in small orchards and 16.00 per cent in medium orchards. The benefit cost ratio was 1:2.47, 1:2.18 and 1:2.39 in marginal, small and medium orchards with pay-back period of 7.5, 7.8 and 7.2, respectively. On an average the net present value of kinnow orchards was ₹ 19823.48 with internal rate of return 15.42 per cent, benefit cost ratio 1:2.40 and pay-back period of 7.6 years.

Table 14 Item wise and concept wise operational costs under different size groups of kinnow orchards (` /hectare)

S.No	Item	Marginal	Small	Medium	Overall
1	Hired Human labour	1777.13	2009.25	1851.63	1830.78
2	Irrigation	0.00	0.00	0.00	0.00
3	Training/ Pruning	232.88	349.10	491.18	265.88
4	Manures + fertilizers	294.05	736.85	487.83	397.83
5	Plant protection	41.50	97.50	89.38	55.28
6	Interest on working capital	281.48	383.13	350.40	305.98
7	Land revenue	0.00	0.00	0.00	0.00
8	Depreciation	308.751	355.63	413.03	322.08
9	EVRL*	3276.18	3373.38	3568.30	3305.90
10	Interest on fixed capital	430.20	447.48	477.75	435.35
11	Family human labour	3125.00	2717.50	2990.00	3030.70
12	cost A (1-6)	2627.03	3575.83	3270.40	2855.73
13	cost B (1-10)	6642.15	7752.30	7729.48	6919.05
14	cost C (1-11)	9767.15	10469.80	10719.48	9949.75

* Earned value of rented land

Table 15 Average returns under different age groups of kinnow orchards (₹/hectare/ year)

Item	**Marginal**	**Small**	**Medium**	**Overall**
Upto 10 years	13815.75	14656.63	15588.88	14051.88
11th to 15th year	16963.30	17385.88	18142.18	17089.95
Above 15th year	17385.80	17813.38	19913.75	17551.05
Overall	16405.28	16935.13	18463.45	16580.18

Table 16 Economic viability under different size groups of kinnow orchards

Group	**Net Present Value(₹)**	**Internal rate of return (%)**	**Payback period (years)**	**Benefit cost Ratio over cost B**
Marginal	18668.83	14.75	7.5	1:2.47
Small	23469.48	15.50	7.8	1:2.18
Medium	29123.08	16.00	7.2	1:2.39
Overall	**19823.48**	**15.42**	**7.6**	**1:2.40**

4

Cultivation, Management and Economic Analysis of Litchi

The litchi (*Litchi chinensis*) is a delicious juicy fruit of excellent quality. It is liked very much as table fruit. Presently, litchi occupies 940 ha area in Jammu region with an annual production of 2315 MT.

Climate and Soil

Litchi being a sub-tropical fruit thrives best under warm humid sub-tropical climate. However, frost in winter and hot winds in summer limits its growth and production. Hot winds results in cracking of fruits before reaching maturity while excessive rainfall at the time of flowering interfere with pollination. The prolonged dry spell after fruit set reduces the crop considerably.

Litchi flourishes well on deep, well drained loamy soil, rich in organic matter and free from hard substrata.

Cultivars

a) **Early** : Muzaffarpur, Early Seedless (Early Bedana)

b) **Mid-Early** : Dehradun, Rose Scented

c) **Late** : Calcuttia, Late Seedless

Propagation

a) **Method :** Litchi is commercially propagated by air-layering (gootee). Non-flowering mature branches may be used for air layering.

b) **Time :** Air-layering of litchi can be done in spring or early summer, however, the period from mid of July to August is considered best for air-layering.

The litchi layers treated with IBA 500 ppm in the first week of August and planted in polythene bags of size 22 x 10 x 8 cm filled with the soil of litchi orchard resulted into 98.29 percent rooting with survival of 91.45 percent within 210 days of planting litchi layers.

Planting

a) **Pit Size :** Before planting, pits of dimension 1m x 1m x 1m should be dug at the desired place a few weeks before the actual planting. The pit should be filled with a mixture of top soil, FYM, which should be mixed with a basket full of soil from a litchi orchard. It is helpful in establishment and quick growth of newly planted plants.

b) **Planting Time :** Planting of litchi should be done when the weather is neither too wet nor too dry. The best planting time is August-September, as the temperature become moderate at that time.

c) **Planting Distance :** Under favourable conditions, litchi is planted at a distance of 10 meter apart both ways i.e., in rows and between plants. Young plants of litchi are very sensitive to heat and frost. They should be protected for first 4-5 years of their orchard life from cold and hot weather. Thatching in winter, frequent irrigations and partial shading in summer can be done to protect the plants from adverse weather conditions.

Training and Pruning

Training of young plants for making a good frame work is necessary. Once desired shape and frame work is achieved, pruning is not required as it happens naturally at the time of fruit harvesting.

Floor Management

a) **Weed control :** Weeds in litchi orchard reduce plant growth and vigour by competing with trees for light, moisture and nutrient. Application of Paraquat and Glyophosate is beneficial in controlling weeds.

b) **Mulching :** Use of mulch is very beneficial in reducing the soil moisture loss and development of better root system of young plants. Suitable fibrous mulches are wheat or barley straw, peanut shells.

Manuring and Fertilization

FYM, diammonium phosphate and murate of potash should be applied in December, half of urea should be applied in the middle of February and other half in the middle of April after fruit set. The doses of FYM and fertilizers according to the age of tree are given in fertilizer schedule as under:

Age of tree (Year)	FYM (Kg/tree)	Fertilizer requirement (g/tree)		
		Urea	Diammonium Phosphate	Muriate of Potash
1	5	95	50	25
2	10	190	105	50
3	15	285	155	75
4	20	380	205	100
5	25	475	260	125
6	30	650	310	145
7	35	830	360	175
8	40	1010	415	200
9	45	1185	465	225
10th year and onwards	50	1490	645	335

Irrigation

Litchi should be carefully irrigated at all stages of development. The irrigation needs are more in summer. The young plants need irrigation twice a week during April-June. In bearing trees, irrigation need is critical during flesh development which occurs from the second week of May to the end of June. Irrigation of litchi trees twice a week during this period reduces cracking of fruits to a great extent and help in proper size development.

Flowering and Fruit set

a) Time of flowering

Litchi trees from air layers may come to flower at the age of 3-5 years, while seedlings take 8-12 years to flower.

b) Fruit set

Though the litchi tree flowers profusely, only a small percentage of the flowers develop into fruits. Several factors such as moisture stress, higher and lower temperatures and strong winds enhance flower drop and reduce fruit set.

Harvesting and Yield

a) Maturity Indices

The maturity of the fruit is determined by the flatness of tubercles and smoothness of epicarp, besides the changes in fruit colour.

b) Method of harvesting

The fruits are harvested in bunches from the tree. While harvesting, the panicles should be cut from the tree with secateurs.

c) Yield

A well cared and fully grown up tree yields 0.8-1.5 quintals of fruit.

Special Problems

Symptoms	Management
Fruit Drop: The premature drop of fruit is thought to be due to failure of fertilization, hormonal imbalance and external factors like high temperature, low humidity and strong winds.	Treatment with growth regulators like NAA @ 20-30 ppm or GA_3 @ 20-50 ppm will minimize fruit drop.Foliar spray of Boron (200 ppm) at the start of anthesis is the most effective treatment in increasing fruit set, yield and quality of Dehradun litchi under sub-tropical agroclimatic conditions of Jammu.
Fruit cracking: Sun burning and skin-cracking of developing fruits are serious problems in litchi and are promoted by high temperature, low humidity and low soil moisture.	Frequent irrigations during critical period of aril growth, spraying of zinc sulphate (1.5%) at weekly intervals, starting from pea size stage of fruit growth to harvest or spraying of GA_3 @ 40 ppm.

Diseases and their management

Symptoms	Management
Leaf Spots **i) Pestalotia Leaf Spot:** (*Pestalotia pauciseta*) Light, discoloured spots on both sides of the leaf appear, having 0.5-2 cm x 0.2 x 2 cm size. Later, the spots coalesce to form bigger lesions, colour of the spot changes from brown to russet.	Spray copper oxychloride (300 g/100 L water), carbendazim (100 g/ 100 L water), captafol (200 g/100 L water) or mancozeb (200 g/100 L water) on the appearance of spots and repeat at 10-15 days interval as per disease severity.
ii) Colletotrichum leaf spot: (*Colletotrichum gloeosporioides*) Spots usually start from the tip of the leaf and extend towards the base, irregular in outline, black to brown in colour with a margin encircling them.	Spray copper oxychloride (300 g/100 L water), carbendazim (100 g/100 L water), captafol (200 g/100 L water) or mancozeb (300 g/100 L water) on the appearance of spots and repeat at 10-15 days interval as per disease severity.

iii) Botrydiplodia leaf spot: (*Botryodiplodia theobromae*) Spots usually start from the tip or margin of the leaf lamina, deep chocolate in colour, margin of the spots with irregular outline and brown in colour. Black pycnidial mass appears on both leaf surfaces.	Spray copper oxychloride (300 g/100 L water), carbendazim (100 g/100L water) or bayleton (100 g/100 L water) or mancozeb (300 g/100 L water) on the appearance of spots and repeat at 10-15 days interval as per disease severity.
iv) Microdiplodia leaf spot: (*Microdiplodia litchi*) Margins of the diseased leaves turn yellowish brown to brick red in colour, the coloured area later becomes light brown and exhibits black dotted pycnidia.Coloured area later becomes light brown and exhibits black dotted pycnidia.	Spray copper oxychloride (300 g/100 L water), carbendazim (100 g/100L water), captafol (200 g/100 L water) or mancozeb (300 g/100 L water) on the appearance of spots and repeat at 10-15 days interval as per disease severity.
v) Red Rust: (*Cephaleuros virescens*) Orange yellow to pink, velvety coating of round shape is formed, generally on the lower surface of the leaves. The affected leaves turn dark brown in colour. The whole terminal gets bronzed, velvety and the inflorescence does not emerge from such terminals. The affected leaves become leathery and turn brown.	i) Prune off the affected branches during autumn and rainy season and destroy them by burning. ii) Give three sprays of copper oxychloride (300 g/100 L water) during spring season: first spray during last week of March before buds open, second and third sprays at 10-15 days interval before flowers open. iii) Repeat three sprays during rainy season.

Insect Pest Management

No serious pest has been recorded on litchi. However, unattended orchards are attacked by bark eating caterpillar, for which control measures can be recommended as under mango for the same insect.

Economics Analysis

The information with respect to yearly costs involved in the establishment of litchi orchards with spacing of 10 m × 10 m are as:

Cost of establishment

During the first year of the plantation, orchardists had incurred expenditure on the preparation of land, planting material, digging, filling of pits and planting, irrigation, plant protection, farm yard manure, fertilizers, intercultural operations etc. The operation wise per hectare average costs for the first year is presented in Table 17.The per hectare total costs incurred during first year for the establishment of litchi orchards were found to be ₹ 27635.61. Variable costs per hectare were found to be ₹ 22315.61 with ₹ 6000.00 incurred on digging, filling and planting, ₹ 3000.00 on preparation of land, ₹ 4200.00 on planting material, ₹ 2250.00 on irrigation, ₹ 600.00 on farm yard manure, ₹ 650.00 on plant protection chemicals, and ₹ 2100.00 on intercultural operations whereas fixed costs were equal to ₹ 5320.00 per hectare with ₹ 1500.00 incurred on implements and '3250 as earned value of rented land (EVRL). The per hectare interest on working and fixed capital was worked out to be ₹ 2390.96 and ₹ 570.00, respectively.

Operational costs

The per hectare operational costs consists of yearly expenses on the maintenance of bearing orchard. The item wise and concept wise operational costs for litchi orchard are presented in Table 18. The per hectare variable costs and fixed were found to be ₹ 27809.04 and ₹ 4934.00, respectively with a total of ₹ 32743.04 per hectare. The table further revealed that among the working costs, the highest expenditures of ₹ 6000.00 per hectare were incurred on farm yard manure and lowest of ₹ 1800.00 was found to be incurred as miscellaneous charges. The fixed costs included ₹ 1200.00 on depreciation on machinery and farm inventory, ₹ 3250.00 as earned value of rented land and ₹ 534.00 as interest on fixed capital

Year wise cost and return analysis

The per hectare per year cost and returns from litchi orchards worked out for different age groups are presented in Table 19. The returns started from 5th year of the establishment of the orchard with production of 5 quintals per hectare which increases to 40 quintals per hectare from 10th year and onwards. As far as the net returns are concerned, it was found to be ₹ 33149.50 per hectare in 6th year which increases to ₹ 187256.96 per hectare during 10th year of the establishment of the orchard and onwards. The benefit cost ratio of 1:2.23 was found in 6th year which increases to 1:6.72 during 10th year of the establishment of the orchard which proves the viability of the orchard.

Table 17 Operation wise first year establishment costs of litchi orchards with spacing= 10 m x 10 m (₹/ha) No. of Plants per hectare = 100

S.No.	Item	Cost (in ₹)
1	Preparation of land	3000.00
2	Digging, filling and planting	6000.00
3	Planting material/ Plantation cost	4200.00
4	Irrigation	2250.00
5	Farm Yard Manure	600.00
6	Fertilizers (Urea, DAP, MoP)	124.65
7	Plant Protection Chemicals	650.00
8	Intercultural operations	2100.00
9	Miscellaneous (rope etc)	1000.00
10	Interest on working capital @12 per cent per annum	2390.96
A	Total variable cost (1-10)	22315.61
11	Implements	1500.00
12	Earned value of rented land	3250.00
13	Interest on fixed capital @ 12 per cent per annum	570.00
B	Total fixed cost (11-13)	5320.00
14	Total establishment cost (A+B)	27635.61

Table 18 Item wise and concept wise operational costs of litchi orchards with spacing= 10 m x 10 m (`/ha)

S.No.	Item	Cost (in ₹)
1	Labour	4500.00
2	Intercultural operations and hoeing, weeding	4000.00
3	Irrigation	2150.00
4	Training/ pruning	2500.00
5	Farm Yard Manure	6000.00
6	Fertilizers (Urea, DAP, MoP)	1779.50
7	Plant protection	2500.00
8	Miscellaneous	1400.00
9	Interest on working capital	2979.54
A	Total variable cost (1-9)	27809.04
10	Depreciation	1200.00
11	EVRL*	3250.00
12	Interest on fixed capital	534.00
B	Total fixed cost (10-12)	4934.00
13	Total operational/ maintenance cost (A+B)	32743.04

*Earned value on rented land

Table 19 Year wise cost and return analysis of litchi orchards with spacing= 10 m x 10 m (` /ha)

S.No.	Particulars	Years							
		1	**2**	**3**	**4**	**5**	**6**	**Average 7th to 9th year**	**10th year and onwards**
1.	Total cost	27635.61	17963.15	23075.73	23421.18	24941.14	26850.50	29345.56	32743.04
2.	Production (qtls)	0.00	0.00	0.00	0.00	5.0	12.0	22.0	40.0
3.	Gross returns (in ₹)	0.00	0.00	0.00	-23421.18	25000.00	60000.00	110000.00	220000.00
4.	Net returns (in ₹)	-27635.61	-17963.15	-23075.73	-5921.18	58.86	33149.50	80654.44	187256.96
5.	Benefit cost ratio	0.00	0.00	0.00	1:0.75	1:1	1:2.23	1:3.75	1:6.72

5

Cultivation, Management and Economic Analysis of Ber

Ber (*Zizyphus mauritiana* Lamk.) is one of the ancient and common fruits, indigenous to India. It is growing in wild, semi wild and cultivated forms all over the country in sub-tropical and tropical areas. Area under ber in J&K is 5,376 hectare with annual production of 10,779 metric tonnes.

Climate and Soil

Ber plant is hardy and can grow successfully even under poor soil and unfavourable climatic conditions, where most other fruit trees fail to grow. It prefers a hot and dry climate for its successful cultivation, but needs adequate water during its first fruiting season. It can be grown up to an elevation of 1000 meters above sea level. Excessive atmospheric humidity seems to be a limiting factor in its successful fruiting.

The ber tree develops a deep tap root system with in a short period of its growth and as such adopts itself to wide variety of soils. However, deep sandy-loam soils appear to be better for its plantations. It is well known in its ability to thrive under adverse conditions of salinity, drought and water logging.

Cultivars

i) Sanaur-2 - Mid-season variety, ripening during 2nd fortnight of March and fairly resistant to powdery mildew.

ii) ZG-2 - Less susceptible to powdery mildew.

iii) Nazuk - Less susceptible to powdery mildew.

Propagation and Rootstock

Till recently, the most common method of multiplying the ber plants has been through seeds. For vegetative propagation, the seedling of ber has to be raised for root stock purpose. Generally, seeds of Katha ber (*Zizyphus mauritiana)* are sown 30 cm x 15 cm apart during March-April (after fresh extraction). The treated seeds (GA_3 @ 500 ppm) germinate in 7 to 10 days after sowing. The germination of freshly harvested seeds can be improved by soaking in concentrated sulphuric acid for six minutes and washing them thoroughly with water immediately. One year old seeds stored at room temperature germinate better than that of freshly extracted seeds. Some of the seedlings are fit for T/Shield budding in August-September, while remaining can be budded during March-June.

Top working of wild trees

The wild ber trees (up to 5 years age) are headed back for top working at the height of about one meter from ground level, whereas in case of older trees, limbs should be headed back at about 30 cm from the branch leaving central limb intact, to avoid the risk of drying of the whole tree (this central limbs be cut off when the top worked limbs get established). The heading back operations may be done from January to April. Bordeaux paste (1 kg of copper sulphate + 1 ½ kg of quick lime into 7 ½ ltrs. of water) should be applied to the cut surface to safeguard against diseases/pathogens. The headed back trees produce a number of shoots below the cut end during June-July. Of these, 2-3 well spaced and healthy shoots should be retained and other removed. When these retained shoots become 1.5 cm to 2.0 cm thick, they should be "T" budded with any of the improved cultivar, during June to September. In order to avoid breakage of the budded shoots due to wind hazards, they may be stacked in case of young trees and tied with central limbs in older trees. Within a year, the top worked trees develop a sufficient growth to bear a small crop and in 2nd and 3rd years, these trees start good bearing.

Planting

There are two planting seasons for ber *viz.,* February-March and July-September. The area is laid out and pits of size 1 m x 1 m x 1 m are dug, one month before planting. Pits should be filled up to about 15 cm higher than the ground level by mixing about 20 kg well rotten farm yard manure, 1 kg super phosphate and 100 g urea with soil in each pit. These pits should be flooded with water thereafter, to settle the loose soil. A plant should be set in the centre of a pit, maintaining same level of soil at which it had stood in the nursery. The budding union should remain about 15-20 cm high from the ground level.The plants should be immediately watered after fixing them in the pits. The plants should be irrigated after four to five days at least during the first two months and then once a week for another three to four months or till plant get fully established. It is very important that all the shoots emerging from below the budded point should be immediately removed.

Training and Pruning

In order to develop strong framework of the ber tree, training becomes essential from first year of planting. While training, care should be taken that there should not be any branch below one meter height of the main stem. The total number of main branches may be 4-6. This will enable the tree to grow in good and balanced shape.

Pruning is highly desirable in ber to maintain tree vigour and productivity as well as to improve fruit size and quality. The fruits are borne in the axil of leaves on the young shoots of current season. There fore, annual pruning is necessary to induce healthy shoots, which will provide maximum fruit bearing area on the tree. The best time of pruning is the hot and dry season when plants shed their leaves and go to rest or become dormant. Light pruning i.e. heading back of about one fourth old growth is desirable for getting higher yield of good quality of fruit. Some thinning out of the branches of ber trees is also necessary to avoid too much crowding so as to admit adequate sunlight and facilitate proper aeration.

Manuring and Fertilization

Age of tree (Year)	FYM (Kg/tree)	Fertilizer requirement (g/tree)		
		Urea	Diammonium Phosphate	Muriate of potash
1	10	215	-	-
2	20	435	-	-
3	30	650	-	-
4	40	870	-	-
5 & onwards	50	1085	20	100

It will be desirable to apply total quantity of fertilizers and manures with monsoons in rainfed areas. However, fertilizer application can be splitted into two doses in irrigated belts, first half in July and second half in October.

Irrigation

Ber is grown successfully in arid climate, however, budded trees do need irrigation for better yield. Bearing trees may be irrigated at monthly intervals during the fruiting season i.e. September – October to February- March. When the fruits are about to mature, irrigation may be avoided, otherwise fruits may become too soft and insipid. After the fruit has been harvested, one or two irrigation may be enough, if there is no rainfall.

Control of fruit drop

Application of 20-30 ppm naphthalene acetic acid (NAA), once in the second week of October and again in the second week of November checks fruit drop to the extent of 10 per cent.

Harvesting and yield

The peak season of harvest is from mid-March to mid April, though it could differ somewhat depending upon location, altitude, cultivar etc. As all the fruits do not ripen simultaneously, 3-5 pickings may be required. The average yield of different cultivars at their prime age of bearing vary from 80 to 200 kg fruit per tree. In dry areas, the yield is relatively lower i.e., 50 to 80 kg per tree.

Insect Pest Management

Symptoms	Management
Fruit Fly: Fruit flies damage ripe fruits of Ber. Maggots feed in the fruit pulp, causing rotting.	i) Harvest fruits when they are still hard. ii) Collect and destroy all fallen fruits. iii) Apply Bait sprays in July-August at 10-15 days intervals (Malathion 50 EC 200 ml and Gur 1 kg in 100L water).

Diseases and their Management

Symptoms	Management
Powdery mildew: The disease attacks emerging leaves and fruit at tender stage Symptoms start in form of white powdery mass on leaves or in patches on fruit surface. Spots later turn in to light brown to dark brown discolouration. The infected leaves curl, fruits shrivel and drop down.	Spray tree with dinocap (50g/100L water), sulfex (200g/100L water) or thiophanate methyl (100ml/100L water), repeat sprays at 15 days interval.
Leaf Spots: Spots of various sizes and shapes appear on the foliage. Severe attack may lead to shedding of leaves.	Spray with mancozeb (200g/100L water), copper oxychloride (300g/100L water) or carbendazim (100g/100L water), repeat sprays at 15 days interval as per disease severity.

Economic Analysis

The information with respect to yearly costs involved in the establishment of ber orchards with spacing of 10 m × 10 m are as:

Cost of establishment

During the first year of the plantation, orchardists had incurred expenditure on the preparation of land, planting material, digging, filling of pits and planting, irrigation, training and pruning, plant protection, farm yard manure, fertilizers etc. The operation wise per hectare average costs for the first year is presented in table 23. The per hectare total costs incurred during first year for the establishment of ber orchards were found to be ₹ 26589.36.Variable costs per hectare were found to be ₹ 23229.36 with ₹ 6000.00 incurred on digging, filling and planting, ₹ 4337.95.00 on preparation of land, ₹ 4000.00 on planting material, ₹ 1000.00 on irrigation, ₹ 1200.00 on farm yard manure, ₹ 702.55 on fertilizers, ₹ 500.00 on plant protection chemicals, and ₹ 3000.00 on training/ pruning whereas fixed costs were equal to ₹ 3360.00 per hectare with ₹ 500.00 incurred on depreciation of implements and ₹ 2500 as earned value of rented land (EVRL). The per hectare interest on working and fixed capital was worked out to be ₹ 2488.86 and ₹ 360.00, respectively.

Operational costs

The per hectare operational costs consists of yearly expenses on the maintenance of bearing orchard. The item wise and concept wise operational costs for ber orchard are presented in table 24. The per hectare variable costs and fixed were found to be ₹ 19366.51 and ₹ 3360.00, respectively with a total of ₹ 22726.51 per hectare. The table further revealed that among the working costs, the highest expenditures of ₹ 6000.00 per hectare were incurred on farm yard manure and lowest of ₹ 1000.00 on hoeing and weeding. The fixed costs included ₹ 500.00 on depreciation on machinery and farm inventory, ₹ 2500.00 as earned value of rented land and ₹ 360.00 as interest on fixed capital

Year wise cost and return analysis

The per hectare per year cost and returns from ber orchards worked out for different age groups are presented in Table 25. The returns started from 5th year of the establishment of the orchard with production of 10 quintals per hectare which increases to 60 quintals per hectare from 10th year and onwards. As far as the net returns are concerned, it was found to be ₹ 12237.49 per hectare in 5th year which increases to ₹ 157273.49 per hectare during 10th year of the establishment of the orchard and onwards. The benefit cost ratio of 1:1.69 was found in 5th year which increases to 1:7.92 during 10th year of the establishment of the orchard which proves the viability of the orchard.

Table 23 Operation wise first year establishment costs of ber orchards with spacing 10m×10m (₹/ha) No. of plants = 100 per hectare

S.No.	Item	Cost (in ₹)
1	Preparation of land	4337.95
2	Digging, filling and planting	6000.00
3	Planting material/ Plantation cost	4000.00
4	Irrigation	1000.00
5	Training and Pruning	3000.00
5	Farm Yard Manure	1200.00
6	Fertilizers	702.55
7	Plant Protection Chemicals	500.00
8	Interest on working capital @12 per cent per annum	2488.86
A	Total variable cost (1-8)	23229.36
9	Implements/ Depreciation	500.00
10	Earned value of rented land	2500.00
11	Interest on fixed capital @ 12 per cent per annum	360.00
B	Total fixed cost (9-11)	3360.00
12	Total establishment cost (A+B)	26589.36

Table 24 Item wise and concept wise operational costs of ber orchardswith spacing 10m×10m (₹/ha)

S.No.	Item	Cost (in ₹)
1	Labour	2500.00
2	Irrigation	1200.00
3	Training/ pruning	3000.00
4	Farm Yard Manure	6000.00
5	Fertilizers (Urea, DAP, MoP)	1091.53
6	Plant protection	2500.00
7	Miscellaneous (hoeing, weeding etc)	1000.00
8	Interest on working capital	2074.98
A	Total variable cost (1-9)	19366.51
10	Depreciation	500.00
11	EVRL*	2500.00
12	Interest on fixed capital	360.00
B	Total fixed cost (10-12)	3360.00
13	Total operational/ maintenance cost (A+B)	22726.51

* Earned value on rented land

Table 25 Year wise cost and return analysis of ber orchardswith spacing 10m×10m (₹/ha)

S.No.	Particulars	Years						
		1	2	3	4	5	Average 6th to 9th year	10th year onwards
1.	Total cost	26589.36	12485.70	13966.96	15451.41	17762.51	20920.50	22726.51
2.	Production (qtls)	0.00	0.00	0.00	0.00	10.00	35.00	60.00
3.	Gross returns (in ₹)	0.00	0.00	0.00	0.00	30000.00	105000.00	180000.00
4.	Net returns (in ₹)	-26589.36	-12485.70	-13966.96	-15451.41	12237.49	84079.50	157273.49
5.	Benefit cost ratio	0.00	0.00	0.00	0.00	1:1.69	1:5.02	1:7.92

❑❑❑

6

Cultivation, Management and Economic Analysis of Phalsa

Phalsa (*Grewia subinaequalis*) is one of the hardiest, drought resistant and quick growing fruit crop. Different parts of the plant are used for curing variety of ailments. The ripe phalsa fruits are consumed fresh, in desserts or processed into refreshing fruit and soft drinks enjoyed during hot summers.

Climate and Soil

Phalsa prefers dry conditions and can therefore, withstand drought. It flourishes well under variable climatic conditions, and requires protection from the freezing cold temperature. Also, adequate sunlight and warm or hot temperatures are required for fruit ripening, development of appropriate fruit colour and good eating quality. Kandi areas are, therefore, quite suitable for its cultivation. It can also be grown under irrigated conditions. Phalsa plant grows vigorously under wide range of soils including fine sand, clay or even limestone. However, best results are obtained in well-drained, loamy soils.

Cultivars

There is no distinct cultivar available in phalsa. Some types which have umbrella like plants are called Chhatri as compared to those which are bushy in appearance. The plants of some types are found to be tall and some are found to be dwarf. A type, Sharbati is known in north India, which is believed to be of better quality.

Propagation

It is commercially propagated by seeds. The seeds should be sown in well prepared nursery beds at 4-5 cm apart in rows spaced at 25-30 cm in July-August with the onset of monsoon. However, seeds loose their viability within 3 months when stored at ambient temperature and hence, fresh seeds from fully ripe fruits should only be sown.

Cuttings

Phalsa cuttings can be treated with IBA 300 ppm and planted in last week of July results in 71.00 per cent survival and takes only 14.00 days for sprouting.

Planting

Phalsa can be planted either during July-August or February-March when the plants have shed their leaves. Eight to twelve month old seedlings are better for planting in the field. Usually, planting is done 2.5-3.0 meter apart both ways, thus accommodating about 1,100-1,500 plants per hectare.

Training and Pruning

Phalsa is allowed to develop a bush, so no training is needed. The most important operation in phalsa cultivation is pruning. Pruning is essential for securing high yield of better grade fruits as it bears fruit on current season's growth. Annual pruning to a height of about 0.5 meters encourages fruiting.

Manuring and Fertilization

Being a hardy crop, fertilization is hardly practiced. But since the bearing is on new growth, it would definitely respond to good fertilization. The application of 10-15 kg FYM after pruning followed by 250 g urea per plant after sprouting is optimum for higher yield.The plant is very sensitive to iron deficiency. Therefore, spraying of $FeSO_4$ (0.4%) will be beneficial to the plants.

Irrigation

Phalsa is regarded as a drought resistant plant but occasional irrigation during fruiting season and in dry periods is profitable for growers. The first irrigation is required in the second or third week of February after the application of fertilizer. During March and April, irrigation after every 20-25 days seems to be beneficial. However, during May, irrigation after every 15-20 days is desirable.

Flowering

Flowering in phalsa starts from February-March and continues till May. Flowers are borne in the axil of leaves.

Harvesting and Yield

Phalsa bushes start bearing in the second year. They yield a good commercial crop in the third year after planting. The fruit starts ripening by the end of May and its harvesting continues through June. Harvesting is done by picking. Ripe berries may be picked on alternate days. The average yield from a mature plant ranges from 5-10 kg.

Diseases and their Management

Symptoms	Management
Leaf Spots: (*Alternaria* sp., *Cercospora* sp., *Phyllosticta* sp.) Spots brown to dark brown or with white crust, irregular, small to medium in size, may coalesce to form big, blighted patches on leaves. Some spots are circular, 3-9 mm in diameter with ashy centre.	Spray mancozeb (200g/100L water) or carbendazim (100g/100L water) at 10-15 days interval.
Rust: (*Dasturella grewiae*) Yellow to light brown pustules with uredospores appear on the leaves. Severe infections lead to drying of leaves resulting in defoliation.	Give 2-3 sprays with carboxin (100g/100L water) or metalaxyl (200g/100L water) at 15 days interval.

Economic Analysis

The information with respect to yearly costs involved in the establishment of phalsa orchards with spacing of 2.5 m × 2.5 m are as:

Cost of establishment

During the first year of the plantation, orchardists had incurred expenditure on the preparation of land, planting material, digging, filling of pits and planting, irrigation, pruning, plant protection, farm yard manure, fertilizers etc. The operation wise per hectare average costs for the first year is presented in table 26. The per hectare total costs incurred during first year for the establishment of phalsa orchards were found to be ₹ 64041.60. Variable costs per hectare were found to be ₹ 60177.60 with ₹ 7500.00 incurred on digging, filling and planting, ₹ 3000.00 on preparation of land, ₹ 16000.00 on planting material, ₹ 1700.00 on irrigation, ₹ 19200.00 on farm yard manure, ₹ 2280.00 on fertilizers, ₹ 1000.00 on plant protection chemicals, and ₹ 3000.00 on pruning whereas fixed costs were equal to ₹ 3864.00 per hectare with ₹ 200.00 incurred on depreciation of implements and ₹ 3250.00 as earned value of rented land (EVRL). The per hectare interest on working and fixed capital was worked out to be ₹ 6447.60 and ₹ 414.00, respectively.

Operational costs

The per hectare operational costs consists of yearly expenses on the maintenance of bearing orchard. The item wise and concept wise operational costs for phalsa orchard are presented in table 27. The per hectare variable costs and fixed were found to be ₹ 34664.00 and ₹ 3864.00, respectively with a total of ₹ 38528.00 per hectare. The table further revealed that among the working costs, the highest expenditures of ₹ 19200.00 per hectare were incurred on farm yard manure and lowest of ₹ 600.00 on plant protection. The fixed costs included ₹ 200.00 on depreciation on machinery and farm inventory, ₹ 3250.00 as earned value of rented land and ₹ 414.00 as interest on fixed capital

Year wise cost and return analysis

The per hectare per year cost and returns from phalsa orchards worked out for different age groups are presented in Table 28. The returns starts from 2nd year of the establishment of the orchard with production of 8 quintals per hectare which increases to 32 quintals per hectare from 3rd year and onwards. As far as the net returns are concerned, it was found to be ₹ 49325.00 per hectare in 2nd year which increases to ₹ 281472.00 per hectare during 3rd year of the establishment of the orchard and onwards. The benefit cost ratio of 1:2.61 was found in 2nd year which increases to 1:8.31 during 3rd year of the establishment of the orchard which proves the viability of the orchard.

Table 26 Operation wise first year establishment costs of phalsa orchards with spacing 2.5m × 2.5m (`/ha) No. of plants = 1600

S.No.	Item	Cost (in ₹)
1	Preparation of land	3000.00
2	Digging, filling and planting	7500.00
3	Planting material/ Plantation cost	16000.00
4	Irrigation	1750.00
5	Pruning	3000.00
6	Farm Yard Manure	19200.00
7	Fertilizers	2280.00
8	Plant Protection Chemicals	1000.00
9	Interest on working capital @12 per cent per annum	6447.60
A	Total variable cost (1-9)	60177.60
10	Implements/ Depreciation	200.00
11	Earned value of rented land	3250.00
12	Interest on fixed capital @ 12 per cent per annum	414.00
B	Total fixed cost (10-12)	3864.00
13	Total establishment cost (A+B)	64041.60

Table 27 Item wise and concept wise operational costs of phalsa orchards (` /ha)

S.No.	Item	Cost (in ₹)
1	Labour	3000.00
2	Irrigation	1750.00
3	Pruning	3000.00
4	Farm Yard Manure	19200.00
5	Fertilizers (Urea, DAP, MoP)	2000.00
6	Plant protection	650.00
7	Intercultural operations	1350.00
8	Interest on working capital	3714.00
A	Total variable cost (1-8)	34664.00
9	Depreciation	200.00
10	EVRL*	3250.00
11	Interest on fixed capital	414.00
B	Total fixed cost (10-12)	3864.00
12	Total operational/ maintenance cost (A+B)	38528.00

* Earned value on rented land

Table 28 Year wise cost and return analysis of phalsa orchards with spacing 2.5m × 2.5m (₹/ha)

S.No.	Particulars	Years		
		1	2	3rd and onwards
1.	Total cost	64041.60	30675.00	38528.00
2.	Production (qtls)	0.00	8.00	32.00
3.	Gross returns (in ₹)	0.00	80000.00	320000.00
4.	Net returns (in ₹)	-64041.60	49325.00	281472.00
5.	Benefit cost ratio	0.00	1:2.61	1:8.31

7

Cultivation, Management and Economic Analysis of Pecan

Pecan (*Carya illinoensis*) is a highly nutritive and energy giving nut fruit. It has nutty flavor and superb tree characteristics. Area under pecannut in J&K is 414 hectares with annual production of 86 metric tonnes.

Climate and Soil

Pecan need warm temperate climate and extremes of both temperate and sub-tropical climates are undesirable. It needs adequate chilling hours ranging from 400-600 hours below 7.2°C for different cultivars.Mean monthly temperature ranging from 24.0 to 29.5°C, with little diurnal variation during growing period is best for growth and fruiting. High humidity prevents pollination, increases the incidence of diseases and premature splitting of nuts.

Although pecan nuts are grown in varied soil conditions, but sandy loam soils (pH 5.0-8.0) which are deep, fertile and are rich in organic matter are best for its growth and fruiting. Symbiotic mycorrhiza and fibrous roots permeate the top 1 meter of soil.

Cultivars

The choice of cultivars should be based on the climate, incidence of diseases, chilling hours and number of days required by the different cultivar to mature.

Important commercial cultivars:

i) Western Shelley: It is recommended for high density planting. It is precocious, very prolific bearer and self pollinated.

ii) Desirable: Tree comes into bearing early, a regular and prolific bearer and resistant to scab. Good for high density planting.

iii) Stuart: It gives good yield and comes into bearing late. It should be grown with protandrous cultivar Desirable.

iv. Wichita: It is very prolific bearer and most precocious. Susceptible to scab and has high zinc requirement. The cultivar is protogynous and should be grown with cv.Western Shelley for pollination.

Propagation and Rootstock

The seeds of any available cultivar of pecan are used as rootstock. Stratification of seed at 1-5^{0}C for 70-75 days gives highest germination. Patch budding in August and whip grafting in dormant season have given best results for propagation. High mortality is experienced at transplanting.

Planting

Pecan trees are very large and are planted at a distance of 16-20 m, depending upon cultivar and soil. For precocious cultivars, spacing of 10.7 m x 10.7 m is recommended. Deep ploughing and leveling is done before planting in late winter or early spring. Because of slow growth of fibrous root system, a considerable mortality is experienced at transplanting which can be minimized by removal of top 50% of plant and keeping 50 cm tap root at planting time. In situ grafting of rootstock also gives good results.

Training and Pruning

Training begins with planting when one third to one half of the top is removed. From second year, the plant is trained to central leader system and branches are allowed to emerge as high as 1.5 m to 2.0 m from ground. No narrow angled branch is retained. Yearly pruning of big tree is difficult, however, possible crowded branches should be removed. Since pecan is a terminal bearer, it requires light pruning.

Manuring and Fertilization

Fully grown up pecan trees should be applied with 100 kg FYM every year in December. In addition, apply 500 g NPK mixture (15:15:15) per year age of the tree and 8 kg mixture to the full bearing trees of sixteen years age and above. The fertilizer should be applied in early spring. Pecan are prone to zinc and manganese deficiency which can be prevented with foliar application of 0.5% zinc sulphate and manganese sulphate.

Irrigation

Water requirement of pecans is higher than other nut fruits. Supplemental water treatments significantly increase kernel yield, nut weight and diameter. The common methods of applying water are flooding, furrows, sprinklers or drip irrigation.

Flowering and pollination

After budding or grafting, the pecan trees begin fruit production in a period of 4 to 6 years depending upon the cultivars and vegetative growth. The staminate and pistillate flowers are borne separately on the same tree.

The pecan nut is mostly cross pollinated, and self pollination is very rare. The planting of 5 % pollinizers have been recommended to ensure good fruiting.

Alternate bearing

Alternate bearing is more pronounced in pecan than other nut fruits. A good orchard management involving better nutrition, water management and pest and disease control considerably reduce degree of alternate bearing.

Harvesting and Yield

The indication of colour markings as dots and streaks on the shell is an indication of nut maturity. Generally, the nuts are thrashed down from the trees with bamboo poles or allowed to fall naturally to the ground where from these nuts are gathered manually. Harvesting should be done in dry weather conditions. The commercial production begins from 9th to 10th year, and yield ranges from 20 to 100 kg $tree^{-1}$ from a fully grown tree depending upon cultivar and growth of a tree.

Drying and storage

When pecan lies on wet soil for few days, they discolour and are infected with mould which deteriorates kernel quality. So, they are dried after harvest to bring moisture content to 4.5%. The nuts should be dried at room temperature by constant circulation of low moisture air for about three days.

Diseases and their Management

Symptoms	Management
Leaf Blotch: (*Gnomonia caryac*) Leaves become brownish giving a blotching effect. Circular dead spots appear on nuts. Cankerous symptoms appear on twigs.	Spray with copper oxychloride /mancozeb (300g/100L water) or thiophanate methyl (100g/100L water) at weekly interval until leaves have fully expanded.

Economic Analysis

The information with respect to yearly costs involved in the establishment of pecan orchards with spacing of 10 m × 10 m are as:

Cost of establishment

During the first year of the plantation, orchardists had incurred expenditure on the preparation of land, planting material, digging, filling of pits and planting, irrigation, training, plant protection, farm yard manure, fertilizers etc. The operation wise per hectare average costs for the first year is presented in table 29. The per hectare total costs incurred during first year for the establishment of pecan orchards were found to be ₹ 40619.60. Variable costs per hectare were found to be ₹ 33339.60 with ₹ 6000.00 incurred on digging, filling and planting, ₹ 5000.00 on preparation of land, ₹ 10000.00 on planting material, ₹ 2500.00 on irrigation, ₹ 3600.00 on farm yard manure, ₹ 467.50 on fertilizers, ₹ 1000.00 on plant protection chemicals, and ₹ 600.00 on training whereas fixed costs were equal to ₹ 7280.00 per hectare with ₹ 1500.00 incurred on depreciation of implements and ₹ 5000.00 as earned value of rented land (EVRL). The per hectare interest on working and fixed capital was worked out to be ₹ 3572.10 and ₹ 780.00, respectively.

Operational costs

The per hectare operational costs consists of yearly expenses on the maintenance of bearing orchard. The item wise and concept wise operational costs for pecan orchard are presented in table 30. The per hectare variable costs and fixed were found to be ₹ 34258.56 and ₹ 7280.00, respectively with a total of ₹ 41538.56 per hectare. The table further revealed that among the working costs, the highest expenditures of ₹ 12000.00 per hectare were incurred on farm yard manure and lowest of ₹ 1000.00 on hoeing and weeding. The fixed costs included ₹ 1500.00 on depreciation on machinery and farm inventory, ₹ 5000.00 as earned value of rented land and ₹ 780.00 as interest on fixed capital

Year wise cost and return analysis

The per hectare per year cost and returns from pecan orchards worked out for different age groups are presented in Table 31. The returns starts from 9th year of the establishment of the orchard with average production of 10 quintals per hectare during the year 9th to 11th which increases to 30 quintals per hectare from 15th year and onwards. As far as the net returns are concerned, it was found to be ₹ 163101.50 per hectare in during 9th to 11th year which increases to ₹ 558461.44 per hectare during 15th year of the establishment of the orchard and onwards. The benefit cost ratio of 1:5.42 was found during 9th to 11th year which increases to 1:14.44 during 15th year of the establishment of the orchard which proves the viability of the orchard.

Table 29 Operation wise first year establishment costs of pecan orchards with spacing = 10m x 10m (`/ha) No. of Plants per hectare = 100

S.No.	Item	Cost (in ₹)
1	Preparation of land	5000.00
2	Digging, filling and planting	6000.00
3	Planting material/ Plantation cost	10000.00
4	Irrigation	2500.00
5	Farm Yard Manure	3600.00
6	Fertilizers (Urea, DAP, MoP)	467.50
7	Plant Protection Chemicals	1000.00
8	Intercultural operations (weeding, hoeing etc)	600.00
9	Training	600.00
10	Interest on working capital @12 per cent per annum	3572.10
A	Total variable cost (1-10)	33339.60
11	Implements/ Depreciation	1500.00
12	Earned value of rented land	5000.00
13	Interest on fixed capital @ 12 per cent per annum	780.00
B	Total fixed cost (11-13)	7280.00
14	Total establishment cost (A+B)	40619.60

Table 30 Item wise and concept wise operational costs of pecan orchards with spacing = 10m x 10m (`/ha)

S.No.	Item	Cost (in ₹)
1	Labour	6000.00
2	Intercultural operations and hoeing, weeding	1000.00
3	Irrigation	2500.00
4	Training/ pruning	1200.00
5	Farm Yard Manure	12000.00
6	Fertilizers (Urea, DAP, MoP)	5888.00
7	Plant protection	2000.00
8	Interest on working capital	3670.56
A	Total variable cost (1-8)	34258.56
9	Depreciation	1500.00
10	EVRL*	5000.00
11	Interest on fixed capital	780.00
B	Total fixed cost (9-11)	7280.00
12	Total operational/ maintenance cost (A+B)	41538.56

*Earned value on rented land

Table 31 Year wise cost and return analysis of pecan orchards with spacing = 10 m x 10 m (` /ha)

S.No.	Particulars	Years							
		1	**2**	**3**	**4**	**Average 5th to 8th year**	**Average 9th to 11th year**	**Average 12th to 14th year**	**15th year and onwards**
1.	Total cost	40619.60	23819.60	28375.20	31931.36	34039.52	36898.50	40405.88	41538.56
2.	Production (qtls)	0.00	0.00	0.00	0.00	0.00	10.00	20.00	30.00
3.	Gross returns (in ₹)	0.00	0.00	0.00	0.00	0.00	200000.00	400000.00	600000.00
4.	Net returns (in ₹)	-40619.60	-23819.60	-28375.20	-31931.36	-34039.52	163101.50	359594.12	558461.44
5.	Benefit cost ratio	0.00	0.00	0.00	0.00	0.00	1:5.42	1:9.90	1:14.44

❑❑❑

8

Cultivation, Management and Economic Analysis of Apple

Apple (*Malus* x *domestica* Borkh.) is a premium fruit of the world. It is a typical temperate fruit and main fruit crop of J&K State. The area under this fruit in J&K State is 1,62,971 hectares with an annual production of 17,26,834 metric tones. The acreage under this fruit crop in Jammu Division is 18,146 hectares and the annual production is 38,421 metric tones. In Jammu Division, apple is being grown in upper areas of Doda, Kishtwar, Ramban, Poonch, Udhampur, Reasi, Kathua and Rajouri districts.

Climate and soil

Apple is a typical temperate fruit crop requiring 1000-1600 hours of rest period below 7°C. The average temperature during growing season should be 21°C to 24°C.Sunshine is essential for colour development of the fruit. In general, a cool climate with low winter temperature and little rainfall in summer is most suitable for apple cultivation.

Apple is grown on a wide range of soils. However, a well drained slightly acidic (about 6.5 -6.7 pH) loam soil with good depth (more than 45 cm) is ideal for apple cultivation. Apple can't tolerate wet field, and water logging may cause collar rot and death of trees.

Cultivars

Most suitable cultivars are those which have commercial value and are also effective for cross pollination. Some of the important cultivars grown in the J&K state are:

Early season cultivars : Mollies Delicious, Starkrimson, Red June, Irish Peach, Benoni and Tydeman's Early Worcestor (P*)

Mid season cultivars: American Mother, Razakwar Jonathan (P*), Cox's Orange Pippin (P*), Red Gold, Queen's Apple. Rome Beauty and Scarlet Siberian.

Late season cultivars : American Apirouge, Ambri, King of Pippins, Lal Ambri, Sunahari, Chamure, Red Delicious, Golden Delicious (P*), Baldwin, Yellow Newton (P*) and White Dotted Red.

*P – These varieties can be used as pollinizers also.

Scab resistant cultivars : Firdous and Shireen have been released for commercial cultivation in J&K State.

Low chilling Cultivars : Vered, Michael, Anna, Tamma, Neomi (less than 800 hours chilling needed). Tropical Beauty and Parlin's Beauty are suitable for processing.

Propagation and Rootstock

The apple plantations are raised on seedling rootstock. Seeds of Crab apple, White Dotted Red, Yellow Newton, Golden Delicious and Northern Spy have good seed viability, germinability and seedling growth. The treated seeds (as mentioned in the raising of nursery chapter) are placed between 2-3 cm thick layer of moist sand in wooden boxes or polythene bags in December. These boxes or bags are kept in cool places having 1-5^{0}C temperature for 50 to 70 days. However, the stratification requirement of seeds is also met in areas of very cool winters by direct sowing of seeds in nursery beds in November-December.

Clonal rootstocks

The size controlling clonal rootstock of Malling (M) and Malling Merton (MM) can be used in deep, fertile and plain soils having assured irrigation facilities. Use of dwarfing rootstocks will result in early bearing and increase productivity per unit area of land.

Clonal rootstock for apple

Category	Root Stock	Characteristics
Dwarfing	M9, M26	Short Juvenile phase, weak anchorage suitable for high density plantation in irrigated areas.
Semi-dwarf	M4, M7, MM106	Susceptible to collar rot, suitable for high density plantation and well drained soil.
Semi-vigorous	M2, MM 111	Tree size 70% of standard, drought tolerant, MM111 is resistant to wooly aphids

The common method of propagation of clonal rootstocks is by mound layering (stooling). The desired cultivars can be grafted or budded on seedling or clonal rootstocks.

Methods of grafting and budding

Method of grafting	:	Tongue grafting
Time of grafting	:	February-March
Method of budding	:	T-budding
Time of budding	:	July-August

The bud sticks should be taken from trees which are healthy, producing excellent fruits of superb quality and prolific and regular bearers. The bud sticks collected prior to grafting should be kept slightly moist with moist saw dust or moss and stored at low temperature to prevent bud break. The bud sticks may also be stored at 5-6^{0}C in wrapped bundles till the period of grafting.

Planting

a) Planting System	:	Square
b) Planting Distance	:	6 m - 7.5 m
c) Planting Time	:	March

In high density plantation on clonal rootstocks, the number of plants per hectare can vary from 500 to 1666 plants per hectare with a planting distance of 5 m x 4 m and 3 m x 2 m, respectively.

The grafted/budded plants should be planted in December in the previously dug pits of 1 m x 1 m x 1 m size. During planting, the prime consideration should be the planting of sufficient pollinizers to ensure effective pollination.

Pollination

Most of the apple cultivars require adequate cross-pollination for optimal yields. For delicious group of apple, 20-25% pollinizers are adequate for getting a commercial yield, however, their proportion can be increased to 30-50% if the plantations are affected by adverse climatic conditions. Early, mid and late season pollinizers provide assured cross-pollination to main cultivar. At some places pollinizers like Golden Delicious and Red Gold show biennial bearing tendency and lead to crop failure in off year, hence new pollinizers like Summer Queen, Tydeman's Early Worcester, Lord Lambourne, Granny Smith, Spartan,

Glauster and Red McIntosh need to be diversified as pollinizer group. The placement of 6-8 bee hives/ha is highly beneficial to increase yield in the orchard lacking in proper proportion of pollinizers.

Training

Training in apple trees is done for giving desired shape to a tree during early years.The apples are trained to a modified leader system.The plant is pruned to 90 cm above the ground immediately after planting. In subsequent years, 3-4 scaffold limbs around the main trunk are developed in different directions.It is not necessary to have all scaffold limbs in first year only, they can be developed in subsequent years. Weak branches and branches with narrow crotch angles should be cut back. Rope and pegs can be used to make crotch angles wider, to induce early flowering.

Pruning

a) Pruning of non-bearing trees

After 3-4 scaffold limbs are developed, very light and corrective pruning is done. Thinning is done to space laterals, narrow and cris-crossed branches are removed.

b) Pruning bearing trees

Apples bear mostly on spurs. Pruning should be done to maintain balance between vegetative growth and spur development. For this, light to moderate pruning should be done in bearing trees. Moderate thinning and heading back should also be done. Over crowding branches should be removed. Cut should be made close to branches leaving no stubs. Pruning should start from top of tree to down wards. Upward growing branches should be cut back to strong laterals. Remove dead and diseased wood. Competing branches should be thinned out. Large cuts should be covered with Bordeaux paste.

Fertilization

Age of tree (Year)	Fertilizer requirement (g/tree)		
	Urea	Diammonium Phosphate	Muriate of potash
1	35	20	50
2	70	45	100
3	105	65	150
4	140	85	200
5	175	110	250
6	245	150	350
7	315	195	450
8	385	235	530
9	455	280	630
10	525	325	730
11	630	385	900
12	735	450	1050
13	840	515	1190
14	940	580	1345
15th year & onwards	1050	645	1500

Note: In matured trees, di-ammonium phosphate and murate of potash should be applied in December. Urea should be applied in 2 split doses. 1st dose should be applied in first fortnight before expected bloom. Second dose may be applied about 3 weeks after fruit set.

Besides, the following schedule of nutrients is recommended for correcting their deficiencies.

Element	Formulation	Concentration (%)	Time of spray
Nitrogen (N)	Urea	0.5	After petal fall
		5.0	Pre-fall in Oct.-Nov.
Calcium (Ca)	$CaCl_2$	0.5	30-45 days before harvest
Zinc (Zn)	$ZnSO_4$	0.5	After petal fall
Manganese (Mn)	$MnSO_4$	0.5	After petal fall
Boron (B)	H_3BO_3	0.1	At pink bud and after petal fall

Irrigation

In case there is a prolonged dry weather, 2-3 irrigations should be given in June, July and August. Areas with excessive rainfall or low lying should be provided with drainage channels to drain out excess of water.

In rainfed areas, mulching with 15 cm layer of FYM/straw/cut grass can conserve moisture and improve soil conditions.

Orchard soil management

Dry grass mulching (10 cm thick) with prior application of glyphosate (800 ml/h) is effective for weed control and moisture conservation. Three grasses *viz.*, red clover (*Trifolium pratens*), fescue (Festuca pratensis) and orchard grass (*Dactilis glomerata*) can be grown in the orchard to increase soil fertility and for the use as a nutritional fodder for milch animals.

Flowering and fruiting

Application of 3% dormex (hydrogen cynamide), 40 days before bud break enhances flowering and improves fruit set. Promalin (GA + 6 BA) @ 30-60 ppm spray at king bloom stage is effective in improving oblong shape of Delicious apples which is a desirable characteristic for better marketing.

Fruit thinning

Heavy bearing in apple not only results in small sized poor quality fruits but also sets in alternate bearing cycle. The judicious thinning (30-40 leaves/fruit) suitably used at the proper stage of fruit development can regulate cropping and improve fruit size and quality.Carbaryl 750-1000 ppm at petal fall to one month after petal fall is effective for fruit thinning.

Harvesting and yield

The apple should be harvested at proper picking maturity. Some of the maturity indices are:

i) When apples attain proper size peculiar to variety.

ii) The fruit should attain intense skin colour.

iii) In most apple cultivars seed colour should be brown.

iv) The calendar dates for harvesting a particular cultivar in a particular locality may vary from 5-10 days over a period of year.

v) Bloom days: Each cultivar requires specific number of days from full bloom to maturity. They may vary slightly due to weather conditions.

The farmers should rely on their experience for picking and also keeping above indices in view. But before starting harvesting, orchard ladders, trained persons, packing material etc. should be kept ready. For removing field heat, the fruit should be precooled (hydrocooled or air cooled) before placing in cold storage. The yield of apple varies with rootstocks, cultivars, soil fertility, weather conditions and orchard management practices. However, on an average, an apple tree yields 30 to 50 kg fruit per year.

Special problems

Symptoms	Management
June drop: June drop caused by moisture stress and hostile environment	It can be reduced by maintaining soil moisture through irrigation and mulching.
Pre-harvest drop: Most serious economic loss caused by reduction of auxin level and increase in ethylene level.	The pre-harvest drop can be checked by application of 10 ppm Naphthalene acetic acid (NAA) before the expected fruit drop or 20-25 days before harvest.
Irregular bearing: The apple cultivars Red Delicious, Royal Delicious and Golden Delicious and some other cultivars have tendency to bear a heavy crop in one year followed by a poor crop next year.	The alternate bearing habit can be checked to some extent by i) Thinning of flowers in oneyear with 10-15ppm NAA at full bloom. ii) Nitrogen application. iii) Judicious spur pruning in winter to decrease crop load. iv) Application of growth retardants like paclobutrazol @ 500-1000 ppm in June – July.
Bitter pit: A physiological storage disorder caused by calcium deficiency is charact-erized by breakdown of cells under skin causing slight depressions generally concentrated at the calyx end of the fruit. Affected tissues become dry and spongy with bitter taste. Large fruits and light crop, excessive nitrogen use and fluctuating soil moisture may result in bitter pit.	Fruit dipping in 4.0 % $CaCl_2$ for 30 seconds before storage.

Water core: Affected fruits become spongy and having dull skin, water soaked areas in the flesh and fermented taste.	Proper calcium nutrition and judicious use of boron are effective.
Cork spot: Affected fruits have small blushed areas on skin and brown spots may be anywhere in the cortex. Affected fruits have harder tissues than normal fruits.	Proper boron nutrition should be followed.

Diseases and their Management

Symptoms	Management
Apple scab: (*Venturia inaequalis*) The disease symptoms are most conspicuous on leaves and fruits. Minute, lesions develop on the lower surface of freshly emerged leaves. These lesions are olivaceous velvety and turn brown to black with age and do not have definite margin. On young leaves these lesions have a radiating appearance with feathery edges. Lesions on older leaves form a convex surface on the opposite side. In heavy infection leaf blades may be curled, dwarfed and distorted. On young fruits, the lesions are small and dark in colour which soon turn almost black. Small, superficial or large patches develop on the fruit surface. Severe early attack leads to formation of mis-shapen fruits that may cause shriveling and cracking. Lesions on twigs are small, 3-5 cm long and slightly raised which become blistered and cause bursting of bark.	i) Spray apple trees with urea (5kg/ 100 L water) at pre-leaf fall. ii) Give one chemical spray before initiation of leaf fall with carbendazim (100g/100L water) or thiophanate methyl (75 g/100 L water) iii) Clean cultivation: Collect and destroy the fallen leaves and pruned, material by burning. Plough the orchard to bury plant debris harbouring the pathogen. iv) Adopt the following spray schedule. **1st Spray:** Silver tip to green tip stage (mid March to mid April): dodine (100 g/ 100 L water), mancozeb (200 g/100 L water), chlorothalonil (400 g/100L water) or dithianon (75g/100 L water) **2nd Spray:** Pink bud stage (1st week of April to end April): hexaconazole (30 ml/100 L water), mancozeb + sulphur (300 g + 200 g), carbendazim (100 g/100 L water) or systhane (40 g/100 L water) **3rd Spray:** Petal fall stage (last week of April to end of May): carbendazim (50 g/100 L water) or thiophanate methyl (50 g/ 100 L water) or hexaconazole (30 ml/ 100 L water)

	4th spray: Fruitlet stage: pea size fruit (mid May to mid June) dithianon (75g/100 L water), dodine (100g/100L water), mancozeb (300g/100L water) or bitertanol (50g/100L water) **5th Spray:** Fruit development stage: Walnut size fruit: 15-21 days after 4th spray (1st week of June to 1st week of July) thiophanate methyl (50g/100L water), diathianon (50g/100L water), dodine (100g/100L water), thiophanate methyl (50g/100L water) or mancozeb (200g/100L water). **6th Spray:** Fruit development stage: 40 days before harvest (mid August) carbendazim (100g/100L water) or bitertanol (50g/100L water) or thiophanate methyl (50g/100L water) or mancozeb (200g/100L water). **7th Spray:** Pre-harvest spray (20-25 days before harvest) carbendazim (100g/100L water)
Powdery Mildew: (*Podosphaera leucotricha*) Small, grayish or white powdery patches appear on young leaves. The infected leaves crinkle and curl up. The grayish white powdery coating covers the entire leaf. Infected foliage becomes hard and brittle. The disease rapidly spreads from one leaf to another. The powdery mass on twigs disappears and brown felt-like covering with dark brown fruiting bodies are seen. The diseased twigs are stunted or get killed and dried up. Floral parts, when affected shrivel and are blighted. Young infected fruits show russeting.	i) Prune infected dormant shoots, tips and shriveled areas to avoid infection in the next season. ii) Spraying the trees at green tip, petal fall and 20 and 40 days after fruitlet stage with dinocap (50g/100L water), carbendazim (50g/100L water), tridemorph (100g/100L water), carboxin (100g/100L water) or fosetyl-al (70ml/100L water)
Sooty Blotch and Fly Speck: (*Sooty* Blotch: *Gloeodes pomigena*: Fly speck:	i) The best way to control these diseases is to carry on pre-harvest

Microthyriella rubi) The twin diseases appear superficially on the skin of apple fruit together and are mistaken as a single disease. The disease becomes severe in late summer with the onset of rains. Sooty blotch is characterized by dull-black to grey spots with indefinite outline. Fly speck lesions are definite, circular, black and often glistening spots resembling sooty blotch in size and colour. These are not deep seated on the fruit skin, but their presence reduces the market value of the fruit. These diseases may express themselves during transit/ storage.

sprays. Give 1st spray 40 days before harvest and the second spray 15-20 days before harvest with the following fungicides captafol (200g/100L water) or carbendazim (100g/100L water) or mancozeb (300g/100L water) ii) Dip the harvested fruit for 1 minute in stable bleaching powder (5kg/100Lwater) or sodium chlorate (3kg/100L water) before packing. iii) Pruning for aeration and sunlight.

Collar Rot: (*Phytophthora cactorum*) The disease appears mostly near the graft union or on the lower trunk or at pruning wounds. The bark of the diseased tree at soil level becomes cankered, soft and spongy. The necrotic tissue turns dark brown and wood beneath the bark is stained dark brown. Bark above ground dries out and splits from the wood. Leaves become purple red in late summer and fall. The affected tree shows little or no shoot growth and girdling leads to death of the tree.

i) Clean the infected collar area with the sharp knife and apply Bordeaux paste or Chaubatia paste or copper oxycholoride paste. ii) Raise the bud/ graft union height to 40-70 cm from soil level. iii) Pre-plant dipping of seedlings in fungicides like difolatan (200g/100L water). iv) Adopt cultural practices like removal of crop residue and fallen fruits, avoid injury to stem, removal of weeds and long grasses from the tree basin.

White Root Rot: (*Dematophora necatrix*) The disease attacks roots of the plant which gives reflection of symptoms on the foliage. Rotting of fine roots is observed which extends to secondary and tertiary roots and ultimately to tree trunk. In advanced stages, roots are completely eaten up by the fungus and seedling/tree is easily uprooted. The symptoms on the foliage are premature yellowing, leaves and fruits remain small in size and cessation of shoot growth. White fungal growth is seen entangled in the root system. In advanced stage

i) Pits in the known infested soil be treated with formaldehyde (50g/L water) before new plantation.
ii) Rotten roots should be cut and the cut ends painted with disinfectant paste (Bordeaux paste or Chaubatia paste). iii) Infested soil of basins be replaced or treated separately with formaldehyde (50g/L water) during dormancy. iv) Central drainage system should be followed. v) Apply carbendazim (100g/100L water), metalaxyl (100g/100L water), fosetyl-al (100ml/100L water) in deep holes

when root system is shattered, the tree dies.	made in the tree basin 20-40 cm apart with a crowbar. Repeat the treatment 3-4 times every year during rainy season or simply drench the seedlings.
Black Rot and Canker: (*Physalospora obtuse*) The disease occurs in three phases *viz.* leaf spots (frog eye spot), fruit rot (black rot) and canker. The most serious is the development or cankers on limbs, trunks andbranches. On large limbs, cankers usually develop on upper surface as reddish brown, sunken lesions which are initially smaller, but slowly increase in size. These turn smoky and show concentric rings. The bark beneath the rough exterior becomes dry, hard and tough. Cankers develop lengthwise more rapidly and become elliptical on limbs and branches and may extend to more than one meter in length and may completely girdle the affected part. Sometimes, a canker represents a superficial roughening of bark. As the infection advances deep into wood, it cracks and falls away exposing the slit wood which is stained reddish brown.	i) Avoid mechanical injury. Apply wound dresser on the wounded areas (Bordeaux paste). ii) Remove girdled limbs, cankers must be cleaned with sharp knife and covered with Bordeaux paste or Chaubatia paste or copper oxychloride paste. iii) Remove mummified fruits, dead and pruned twigs from the orchard and burn. iv) Spray the trees with carbendazim (100g/100L water) or thiophanate methyl (100g/100L water).
Stem Brown: (*Botryosphaeria ribis*) Cankers and fruit rot are two phases of the disease. The cankers usually appear in summer months as brown, depressed lesions starting from pruned surface or around the insect holes. Sometimes, blisters appear on the bark which exudes watery liquid spreading over the surface of the lesions. Up to the following spring, a few pimple like eruptions develop on the smooth bark of the lesion as a result of production of fruiting bodies underneath the bark. The small cankered lesion show die-back and become wrinkled. The bark is loosened. It becomes brown and papery. Wood below the bark is	The disease can be effectively controlled by adopting judicious pruning of all dead wood and proper management. Normal fungicidal spray schedule, as recommended for apple scab, if properly followed, takes care of the disease. Spray any of the fungicide recommended in that schedule at 10-15 days interval as per severity of the disease.

stained dark brown. It turns slimy and develops cracks. An alcoholic smell is emitted. Large limbs may be girdled by fusion of cankers.The symptoms on fruits are irregular, light green spots encircled with brown halo on the yellow skin of the young fruits. Spots later become margins. Some of these spots produce a series of concentric rings as well. Infection on the mature fruit may develop rapidly under warm conditions and the entire fruit is decayed.	
Apple Mosaic: (*Apple mosaic virus*) Plants show variable symptoms in the field. Creamy white or yellow patches appear on the leaf lamina. A part or the whole leaf turns yellow due to numerous spots. Leaves become necrotic and wither away. Plant loses vigour gradually and yield is reduced considerably. The virus is transmitted through bud-wood by natural rootgrafting and budding.	i) The bud wood and grafted seedlings can be rendered virus free by keeping at 36^{0}C for 4 weeks. ii) Keep the infected bud wood at 50^{0}C for 8 minutes before budding. iii) Maintain seedlings at 37^{0}C for 27 days or at 40^{0}C for 20 days to inactivate the virus. iv) Virus free mother plants be indexed for graft and bud-wood material.
Pre Mature Leaf Fall: (*Marsoninna mali*) Dark Brownspots of 5-10 mm appear on mature leaves in spring or early summer which turn yellow and drop within few weeks. Small dot like structures appear on such lesions.	i) Collect and burn fallen leaves in autumn. ii) Spray tree with dithianon (50g/100L water), thiophanate methyl (50g/100l water) or dodine (75g/100L water). Give 4-5 sprays at 20 days interval from fruit set up to harvesting.

Insect Pest Management

Symptoms	Management
San Jose Scale: (*Quadraspidiotus perniciosus*) All parts above ground are attacked. Heavily infested trees lack vigour, its foliage turn thin and yellow speckled. In the beginning only growth is checked but as the infestation increases, the attacked trees die due to devitalization. More often certain limbs become infested and die before the whole tree is affected. Even after successful control of scale on heavily infested trees/limbs, the branches and buds may be so weakened that no fruit is borne for one or two seasons. Often bark around scale is reddened. On fruits the insects appear as grey patches surrounded by inflamed red area.	Spray orchard during dormancy with 3% tree spray oil or diesel oil emulsion prepared locally with fish oil soap at 6.33% i.e., stock solution and water in 1:5 ratio. The dormant spraying should be done after leaf fall in clear weather and temperature is about 4^0C. Spray with high volume sprayer to cover the plant surface thoroughly. Where spraying during dormancy has missed it may be done at late dormant stage (upto ½" green tip stage) and in that event the diesel oil emulsion may be diluted 10-15 times and tree spray oil 2-2.5% depending upon the stage of foliage development. In case of late dormancy, spraying the tree spray oil may be impregnated with either chlorpyriphos 20EC (0.02%), or methyl demeton 25EC (0.025%) to help in suppression of Sanjose Scale and reduction in population of European red mite, aphids, thrips and other infestation in early spraying.The spray should be applied thoroughly to all parts of the tree. In case of rains washing the spraying of oil emulsion, spray should be repeated. The normal orchard practice of dormant spraying, but heavy encrustation often requires additional applications to control crawlers. At 80% of petal fall of Red Delicious variety (as indicator variety) spray the trees with any of the insecticides, chlorpyriphos (0.02%), quinalphos (0.025%), methyl demeton (0.025%) against adult male of San Jose Scales. The time of application is most important to control the heavy scale infestations. The treatment of

	petal fall will also suppress the population of aphids, mites, leaf miners, tortricid moths and other pests. The spraying may be repeated in the first week of June against crawlers of first generation. If needed third spraying may be done in the last week of August when the crawlers of second generation start emerging. Care should be taken to repeat insecticides during summer control programme. If the crop is ready for harvest the trees may be sprayed with dichlorvos (0.05%). The fruit should not be harvested before waiting period.
Apple Aphid: Feeding on plant results in severe distortion of leaves which may turn yellow or brown resulting in premature fall. Infested trees become sick and have stunted growth.	Dormant spray as for San Jose Scale gives adequate control. However, spray with methyl demeton 25 EC (0.025%), chlorpyriphos 20 EC (0.02%), dimethoate 30 EC (0.03%), malathion 50EC (0.05%) in late May or early June on young trees.
Woolly Aphid: (*Eriosoma lanigerum*): Woolly aphids suck the sap of branches, twigs and roots. Galls are often formed at the point where aphids feed. Infested plants become sickly and stunted in growth.	i) In addition to dormant sprays for San Jose Scale, application of methyl demeton 25 EC (0.025%), dimethoate 30 EC (0.03%), malathoin 50 EC (0.05%) between petal fall and June will prevent infestation during summer. Serious infestation near the harvest can be controlled by using dichlorvos 76 EC (0.05%) during summer. ii) Infestation on non bearing young trees (1-4 years old) can be reduced by placing carbofuran 3G 30g granules at 5 cm depth in the root zone in April-May when there is sufficient moisture in soil. In case no moisture is available give light irrigation.
Totricid Moth / Bud caterpillar: (*Cacoecia sarcostega*): Larvae often feed on leaves buds and small fruits. They often spin a light web around several	Same as in San Jose Scale. Spray phosphamidon or chlorpyriphos after petal fall.

leaves which they roll or draw together often enclosing cluster of apples. They may eat cavities in centre or sides of the fruits.

Apple Leaf Miner:(*Gracilaria zachrysa*) Larvae mine the leaves. Tissue surrounding the mines become discoloured. If several mines are present, infested leaves may eventually shrivel, die and all.

Spray at green tip stage with tree spray oil against San Jose Scale will give effective control. Application of methyl demeton 25EC (0.025%), dimethoate 30 EC (0.03%) after petal fall will suppress the pest.

Hairy Caterpillar: (*Euproctis* sp.) Caterpillars feed on leaves. In heavy infestation, trees are completely defoliated. Repeated defoliation may weaken and kill most of the trees.

Hairy caterpillar may be managed by:
i) Collection and destruction of egg masses.
ii) Stapling burlap skirts around tree trunks infested with caterpillar and collection and destruction of larvae/ pupae from May to the end of June.
iii) Spraying after petal fall chlorpyriphos 20 EC (0.02%) or quinalphos (0.02%)

Shot/Pin Hole Borer: (*Scolytoplatypusraja*) Adult and grubs tunnel into the sapwood and hardwood of the plant making galleries and pin holes. Surface of infested branches get perforated followed by yellowing and wilting of leaves. A serious infestation may kill the whole trees. Holes are some times indicated on the bark.

Shot/pin-hole borer may be controlled by:
i) Pruning and destruction of borer infested branches.
ii) Up-keep the vigor of trees by application of balanced dose of fertilizers will help in reducing the attack of shot hole borer/stem borer.
iii) Balanced use of fertilizers.
iv) Leave scattered, fallen tree trunks and branches in the orchard for trapping egg laying of scolytids during the month of June or first week of July.
v) Swabbing infested branches and trunks with 1.5% lindane dust and soil in the ratio of 1:6 mixed with rice wheat straw.

Stem Borer: (*Apriona cinerea*) The beetles damage the stem and branches drilling big holes in the trunk. Saw dust is seen coming out from holes.

To avoid serious damage by stem borer, clean the hole and plug it with cotton soaked in petrol/dichlorovos after inserting naphthalene balls. Swabbing as in Shoot/Pin hole borer.

Chaffer Beetles: (*Adoretus* spp.) Adult beetles feed onleaves, buds, blossom and fruit/lets. Eaten away leaves are perforated. Thegrubs feed on roots andmay cause wilting of the plant.	i) Spray carbaryl 50 WP (0.1%) during May-June at dusk. ii) Shaking of young non-bearing trees at dusk and collecting and destroying the beetles is useful. iii) Apply carbofuran 3G (1kg a.i./ha) in soil against grubs.
European Red Mite (*Panonychus ulmi*) **Two spotted spider mite:** (*Tetranychus urticae*) The mites puncture the tissues of the leaves and feed on plant juice. If the infestation is light the foliage becomes speckled heavily infested leaves become yellowish or bronzed. Severe infestation will result in early dropping of leaves, retarted growth, weakened fruit buds and reduced size of fruits.	Dormant oil emulsion spray or delayed dormant spraying with mixture of tree spray oil and organophorous compounds will provide adequate control of European red mite. If 5-8 mites/leaf are observed during June spray of abamectin (500ml/ha) or dicofol (100ml/100 L water) or methyl demeton (80ml/100 L water).Certain fungicides like wettable sulphur, mancozeb also reduce the mite population.
Bud and flower thrips: (*Thrips* spp.) Thrips feeding within the buds may cause weeping of sap from injured areas. Later brownish patches are produced at the bases of stamens and styles and on petal of infested blossom. Attacked leaf tissue also becomes distorted. Injury may led to distortion of petals, resetting of fruits and reduction of fruits set.	Control measures recommended for San Jose Scale shall also take care of blossom thrips.

Economic Analysis

The information with respect to yearly costs involved in the establishment of apple orchards with spacing of 6 m × 6 m are as:

Cost of establishment

During the first year of the plantation, orchardists had incurred expenditure on the preparation of land, planting material, digging, filling of pits and planting, irrigation, training, plant protection, farm yard manure, fertilizers etc. The operation wise per hectare average costs for the first year is presented in table 32. The per hectare total costs incurred during first year for the establishment of apple orchards were found to be ₹ 48483.06. Variable costs per hectare were found to be ₹ 40643.06 with ₹ 5560.00 incurred on digging, filling and planting, ₹ 6950.00 on preparation of land, ₹ 16680.00 on planting material, ₹ 1500.00 on irrigation, ₹ 2780.00 on farm yard manure, ₹ 318.45 on fertilizers, ₹ 1000.00 on plant protection chemicals, and ₹ 600.00 on training whereas fixed costs were equal to ₹ 7840.00 per hectare with ₹ 2000.00 incurred on depreciation of implements and ₹ 5000.00 as earned value of rented land (EVRL). The per hectare interest on working and fixed capital was worked out to be ₹ 4354.61 and ₹ 840.00, respectively.

Operational costs

The per hectare operational costs consists of yearly expenses on the maintenance of bearing orchard. The item wise and concept wise operational costs for apple orchards are presented in table 33. The per hectare variable costs and fixed were found to be ₹ 38559.28 and ₹ 7840.00, respectively with a total of ₹ 46399.28 per hectare. The table further revealed that among the working costs, the highest expenditures of ₹ 12000.00 per hectare were incurred on farm yard manure and lowest of ₹ 1200.00 on irrigation. The fixed costs included ₹ 2000.00 on depreciation on machinery and farm inventory, ₹ 5000.00 as earned value of rented land and ₹ 840.00 as interest on fixed capital

Year wise cost and return analysis

The per hectare per year cost and returns from apple orchards worked out for different age groups are presented in Table 34. The returns starts from 8th year of the establishment of the orchard with average production of 41,70 quintals per hectare during the year 6 to 10 which increases to 112 quintals per hectare from 15th year and onwards. As far as the net returns are concerned, it was found to be ₹ 171086.43 per hectare in during 6th to 10th year which increases to ₹ 513600.72 per hectare during 15th year of the establishment of the orchard and onwards. The benefit cost ratio of 1:5.57 was found during 6th to 10th year which increases to 1:12.07 during 15th year of the establishment of the orchard which proves the viability of the orchard.

Table 32 Operation wise first year establishment costs of apple orchards with spacing = 6m x 6m (`/ha) No. of Plants per hectare = 278

S.No.	Item	Cost (in ₹)
1	Preparation of land	6950.00
2	Digging, filling and planting	5560.00
3	Planting material/ Plantation cost	16680.00
4	Irrigation	1500.00
5	Farm Yard Manure	2780.00
6	Fertilizers (Urea, DAP, MoP)	318.45
7	Plant Protection Chemicals	1000.00
8	Intercultural operations (weeding, hoeing etc)	900.00
9	Training	600.00
10	Interest on working capital @12 per cent per annum	4354.61
A	Total variable cost (1-10)	40643.06
11	Implements/ Depreciation	2000.00
12	Earned value of rented land	5000.00
13	Interest on fixed capital @ 12 per cent per annum	840.00
B	Total fixed cost (11-13)	7840.00
14	Total establishment cost (A+B)	48483.06

Table 33 Item wise and concept wise operational costs of apple orchards with spacing = 6m x 6m (`/ha)

S.No.	Item	Cost (in ₹)
1	Labour	7200.00
2	Intercultural operations and hoeing, weeding	2000.00
3	Irrigation	1200.00
4	Training/ pruning	2500.00
5	Farm Yard Manure	12000.00
6	Fertilizers (Urea, DAP, MoP)	4527.93
7	Plant protection	5000.00
8	Interest on working capital	4131.35
A	Total variable cost (1-8)	38559.28
9	Depreciation	2000.00
10	EVRL*	5000.00
11	Interest on fixed capital	840.00
B	Total fixed cost (9-11)	7840.00
12	Total operational/ maintenance cost (A+B)	46399.28

* Earned value on rented land

Table 34 Year wise cost and return analysis of apple orchards with spacing= 6m x 6m (` /ha)

S.No.	Particulars	Years							
		1	**2**	**3**	**4**	**5**	**Average 6th to 10th year**	**Average 11th-14th year**	**15th years and onwards**
1.	Total cost	48483.06	28007.73	29921.19	31843.37	33766.17	37413.57	43140.41	46399.28
2.	Production (qtls)	0.00	0.00	7.00	12.20	26.00	41.70	84.00	112.00
3.	Gross returns (in ₹)	0.00	0.00	0.00	0.00	0.00	208500.00	420000.00	560000.00
4.	Net returns (in ₹)	-48483.06	-28007.73	-29921.19	-31843.37	-33766.17	171086.43	376859.59	513600.72
5.	Benefit cost ratio	0.00	0.00	0.00	0.00	0.00	1:5.57	1:9.74	1:12.07

❑❑❑

9

Cultivation, Management and Economic Analysis of Pear

Pear (*Pyrus* spp.) is a superior temperate fruit with good taste and flavour. Because of its tolerance to wide range of soil and climatic conditions, it is grown both in sub-tropical and temperate conditions. In J&K, 14,532 hectare area is under cultivation, with an annual production of 88,329 metric tones. In Jammu Division, pear is being grown on 7,600 hectare having 30,257 metric tonnes production annually. The pear is grown in Doda, Poonch, Rajouri, Udhampur and Kathua districts of Jammu Division.

Climate and Soil

Pear is adapted to wide range of climatic conditions. Most of the commercial cultivars require 900-1000 hours of chilling below 7°C to break the dormancy. Pear is a temperate fruit crop; however, it can be grown successfully in sub-tropical regions of Jammu. There are several low chilling cultivars, which are well adapted for cultivation under sub-tropical conditions. Pears are grown successfully over a wide range of soils provided they have sufficient moisture. Compared to apples, they are less tolerant to drought and rather more tolerant to impeded drainage. Deep, well drained and fertile soil give high yield.

Cultivars

A) For temperate Region

  a) Early season cultivars: Mature from last week of June to third week of July.

    i) China pear (China Tang)

    ii) Beuure-de-Amanils (Zirhami Tang)

  b) Mid season cultivars: Matures in third week of July to third week of August.

    i) Citron-des-Carmes (Goshbugh)

ii) Clapp's Favourite

iii) Doyenne Bossoch (Surajbali Tang)

iv) Fertility (Misri Tang)

v) Williams' Bon Chretien or Bartlett (Awal Number)

vi) Starkrimson

c) Late season cultivars: Matures from last week of August to second week of September.

i) Beurre Hardy (Kharpaddur)

B) For sub-tropical Region

i) Patharnakh: Tree spreading and vigorous, fruit medium round, green, flesh gritty, crispy and juicy, good keeping quality, stands transportation well. Ripens in last week of July. Average yield is 150 kg/tree.

ii) LeConte: Tree medium in vigour and spreading. Fruit small to medium in size and pyriform in shape with alternate greenish yellow colour. Flesh whitish, juicy and sweet. Ripens in August and yields about 60-80 kg/tree.

Some other cultivars recommended for sub-tropical areas are: Kieffer, Punjab Beauty, Punjab Gold, Punjab Nectar etc. Smith is also a good pollinizer for LeConte.

Pollination

Most pears fail to set a good crop with their own pollen and crop better with cross pollination. So, plant two or more cultivars, whose flowering overlaps for a good crop. For adequate pear pollination 5-6 honey bee colonies/ha are optimum at 100-150 m distance. For optimum fruit set and yield in pear, bee hives should be placed at 30-50% flowering, and should be replaced by fresh ones after 3-4 days to increase the number of foraging bees.

Propagation and Rootstocks

The choice of rootstocks lies between pear and quince. Plants grown on quince rootstock can be planted in soils with higher water table and with the object of having dwarf pear trees. However, seedlings of commercial cultivars can also serve the purpose of a good rootstock.

Plants are raised on wild pear (*Pyrus pashia*) stock locally known as Kainth and also on quince seedling. Seedlings of kainth or quinces are raised after stratification of seeds for 30-50 days at 1-5^{0}C.

The stratification can also be met by sowing the seeds in open nursery in November. Seedlings are tongue or splice grafted before the bud break. In temperate areas 'T' budding is commonly practiced in July-August.

Planting

Planting distance depends upon the site, soil, rootstock and cultivars. Generally, the pear trees are planted at the distance of 6-8 m in square system of planting. The planting is done in winter upto middle of February. Generally, one-year-old plants are planted but older plants of 2-3 years age can also be planted with good success. The older plants should preferably be planted by the end of December.

Training and Pruning

a) Pear trees are trained to "modified central leader system". In this, 4-5 well spaced branches are developed in first 3 to 4 years and then leader is headed back.

b) Pruning of young plants differs considerably from that of bearing ones. In case of young plants, unbranched plants should be headed back at 60-90 cm above the ground level at planting time. Pear bears on spurs arising on two year old shoot. In case of old bearing trees, pruning of dead woods and non-bearing spurs should be practiced. Old trees often require spur shortening to maintain size and quality of fruit and vigour of trees. The thinning out and heading back of the laterals may also be done to encourage the fruiting spur formation.

Fertilization

Age of tree (Year)	Fertilizer basis (g/tree)		
	Urea	Diammonium Phosphate	Muriate of Potash
1	35	20	50
2	70	45	100
3	105	65	150
4	140	85	200
5	175	110	250
6	245	150	350
7	315	195	450
8	385	235	530
9	455	280	630
10	525	325	730
11	630	385	900
12	735	450	1050
13	840	515	1190
14	940	580	1345
15th year & onwards	1050	645	1500

Note: Diammonium phosphate and Muriate of potash should be applied as single dose in December, half of urea should be applied 2-3 weeks before expected bloom and another half of urea 3 weeks after fruit set.

Irrigation

Irrigation becomes necessary when there is not enough moisture in soil and there are no summer rains. Excessive irrigation should be avoided.

Orchard floor management

a) Weed control

Weeds can efficiently be controlled by spraying the herbicides namely Paraquat @ 0.6 % or Hexuron or Glycel or Gramaxone.

b) Mulching

Under rainfed conditions, mulching is used to conserve moisture, which also helps in controlling weeds, erosion and temperature fluctuation. Straw mulching is considered better.

Harvesting and Yield

Fully mature pear fruits are harvested while, firm and green for canning and distant market, and for local consumption they are picked at slightly latter stage because fruits remaining on a tree make a considerable gain in size, weight and overall quality. Pear fruits are picked individually by giving slight twist rather than pulling directly. Two to three pickings at an interval of 3-4 days are better than single picking.

The yield of fruit varies greatly with the cultivars, age of the plant, orchard management and climatic conditions. A well managed orchard gives about 30-40 tonnes of fruits/hectare.

Special problems

Physiological disorders	Management
Pre-mature ripening and pink discolouration	Abnormally cool weather before harvesting causes premature ripening and pink colouration near blossom end. Consequently core breakdown (brown heart) and softening occurs in affected fruits which do not ripen properly.
Mis-shapen fruits and corking	They are caused by boron deficiency. To overcome it, 25-40 kg borax should be applied through soil at the time of first dose of fertilizer application after every three years.
Fruit drop	Apply 20 ppm NAA after fruit set for checking early fruit drop. To control pre-harvest drop, application of 10 ppm NAA Three weeks before harvest is beneficial.

Diseases and their Management

Symptoms	Management
Seedling Blight: (*Sclerotium rolfsii*) Infected seedlings are killed and small mustard seed like sclerotia bodies appear at the collar region during monsoon.	i) Uproot and burn the affected seedlings. ii) Drench the nursery soil with combination of carbendazim (12 %) + mancozeb (63 %) @ 100g/100L water or thiophanate methyl (100g/100L water) in nursery soil/before transplantation.
Hairy Root: (*Agrobacterium rhizogens*) Excessive growth of fibrous roots develops from one point and gives broom like appearance.	i) Avoid collar/root injury. ii) Dip healthy grafted plants in 1% copper sulphate solution for 90 minutes. iii)Change the nursery site. iv) Rogue out infected plants.
Crown Gall: (*Agrobacterium tumefaciens*) Globular, elongated or irregular tumors formed at or near the graft union, usually in the nursery plants.	i) Avoid collar/root injury. ii) Dip healthy grafted plants in 1% copper sulphate solution for 90 minutes. iii)Change the nursery site. iv) Rogue out infected plants.
White Root Rot: (*Dematophora necatrix*) Affected seedlings/trees show scanty foliage, slow growth, bronzing and yellowing of leaves. Such plants ultimately die. Roots turn brown and remain covered with white cottony mycelial mat of the fungus during rainy season.	i) Improve orchard drainage. ii) Remove infected roots and apply Chaubatia paint (red lead, copper carbonate and linseed oil 4:4:5) on the cut ends and healthy parts of the roots during November-December iii) Uproot and burn dead plants and treat the soil pit with mercuric chloride solution (1%). iv) For curing the infected trees, give three drenchings of carbendazim (100g/100L water) or chlorathalonil (100g/100L water) at 15 days interval during monsoon. Drenching application should be made in 6-9 inch deep holes at a distance of 30 cm each throughout the canopy of the tree.
Collar Rot: (*Phytophthora cactorum*) Collar region of the tree near the ground level turns brown, soft and spongy and on being girdled, completely kills the tree. The disease spread is highest in poorly drained soils.	i) Remove soil around the collar region in November-December, expose the affected portion to sunlight. Remove the affected bark and apply copper oxychloride paint on healthy portion. ii)Irrigate around the tree trunk in 30 cm radius with metalaxyl (200g/100L water)

	or combined application of metalaxyl + mancozeb (200g/100L water). iii) Keep the graft union 20-25 cm above the soil level. iv) Use M4 and M9 rootstocks as these are resistant to collar rot disease. v) Avoid water-logging conditions in the orchard by staking pebbles or sand in the tree basin.
Canker and Die-Back: (*Pink Canker, Botryobasidium salmonicolor*, European canker, *Nectria galligena*, Black rot or Smoky blight, *Botryosphaeria qercum*, Nail head Canker, *Numularia* discrete, Stem black *Coniothecium chomatosporum*, Silver leaf: *Chondrostereum purpureum*, Stem brown or stem bark canker, *Botryosphaeria dothidea* and other cankers caused by various fungi.	i) Cut and burn the badly cankered portions of the tree. ii) Scrap the cankered portions alongwith somehealthy portion and cover the wounds with Chaubatia paint or copper oxychloride paint or white lead paint. iii) Spray copper oxychloride (300g/100L water) or captafol (200g/100L water) or carbendazim (100g/100L water) after harvest and bud swell stage. iv) Spray schedule adopted for apple/pear scab also helps in reducing the incidence of these cankers.
Pear Scab: (*Venturia pirina*) The Li) Spray pear trees with urea (5kg/100 L symptoms resemble those of apple scab. The disease is more pronounced on *Kashmiri Nakh*. The symptoms are more conspicuous on fruit than on leaves. On fruits, the spots are smaller and darker which soon turn almost black.	i) Spray pear trees with urea (5kg/100 water) at pre-leaf fall. ii) Spray carbendazim (100g/100L water) or thiophanate methyl (75g/100 L water) before initial leaf fall stage. iii) Collect and destroy the fallen leaves and pruned, material by burning. Plough the orchard to bury plant debris harbouring the pathogen.
Leaf Spots: (*Mycosphaerella* spp., *Alternaria mali, A. alternata. Coniothyrium pyrinium, Phyllosticia* spp. and *Botryosphaeria quercum*) Leaf spots of different colour, sizes and shapes are of common occurrence during late summer and rainy season. Severely affected leaves turn yellow and fall prematurely.	i) Spray schedule adopted for pear scab also provides effective control of these diseases. ii) During rainy season, spray 2-3 times a mixture of carbendazim + mancozeb (250 g/ 100 L water) at 15 days interval.
Sooty Blotch and Fly speck: (*Gloeodes pomigena* and *Schizothyrium pomi*) Both the diseases	i) The best way to control these sprays. Give 1st spray 40 days before harvest and the second spray 15-20

appear under high atmospheric humidity in rainy season on the external surface of the fruits and reduce their market value. Dull black or grey spots of sooty blotch and black circular flyspecks appear.	days before harvest with the following fungicides captafol (200g/100L water) or carbendazim (100g/100L water) or ii) Dip the harvested fruit for 1 minute in stable bleaching powder (5kg/100L water) or sodium chlorate (3kg/100L water) before packing (iii) Pruning for aeration and sunlight.
Powdery mildew: (*Podosphaera leucotricha*) Symptoms resemble those in apple. The disease affects the growth of buds, new shoots and leaves, and produces white powdery patches.	i) Prune infected dormant shoots, tips and shriveled areas to avoid infection in the next season. ii) Spraying the trees at green tip, petal fall and 20 and (50g/100L water), carbendazim (50g/100L water), tridemorph (100g/100L water), carboxin (100g/100L water) or fosetyl-al (70ml/100L water)

Insect Pest Management

Symptoms	Management
Sanjose Scale: (*Quadraspidiotus perniciosus*) All parts above ground are attacked, heavily infested trees lack vigour, its foliage turn thin and yellow speckled. In the beginning, only growth is checked but as the infestation increase, the attacked trees die due to devitalization. More often certain limbs become infested and die before the whole tree is affected. Even after successful control of scale on heavily infested trees/limbs the branches and buds may be so weakened that no fruit is borne for one or two seasons. Often bark around scale is reddened. On fruits, the insects appear as grey patches surrounded by inflamed red area.	Spray orchard during dormancy with 3% tree spray oil or diesel oil emulsion prepared locally with fish oil soap (potash base) at 6.33% i.e., stock solution and water in 1:5 ratio. The dormant spraying should be done after leaf fall in clear weather and temperature is about 4^0C. Spray with high volume sprayer to cover thoroughly the plant surface.Where spraying during dormancy has missed, it may be done at late dormant stage (upto ½" green tip stage) and in that event, the diesel oil emulsion may diluted 10-15 times and spray oil 2-2.5% depending upon the stage of foliage development.In case of late dormancy, spraying the tree spray oil may be impregnated with either chlorpyriphos 20 EC (0.02%) or methyl demeton 25 EC (0.025%) helps controlling Sanjose Scale and in suppressing the population of European red mite, aphids, thrips and other infestation in early spraying.The spray

	should be applied thoroughly to all parts of the tree. In case of rains washing the sprarying of oil emulsion spray should be repeated. The light infestations may be controlled by the normal orchard practice of dormant spraying, but heavy encrustation often requires additional applications tocontrol crawlers. At 80 % of petal fall of Red Delicious variety (as indicator variety) spray the trees with any of the insecticides, chlorpyriphos 0.02% quinalphos (0.025 %) against adult male of San Jose scales. The time of application is most important to control the heavy scale infestations. The treatment at petal fall will also suppress the population of aphids, mites, leaf miners, tortricid moths and other pests.The spraying may be repeated in the first week of June against crawlers of first generation. If needed third spraying may be done in the last week of August when the crawlers of second generation start emerging. Care should be taken in repetition of insecticides during summer control programme. If the crop is ready for harvest, the trees may be sprayed with dichlorovos (0.05%). The fruit should not be harvested before waiting period.
Apple Aphid: Feeding on plant results in severe distortion of leaves which may turn yellow or brown resulting in premature fall. Infested trees become sickly and stunted in growth.	Dormant spray as for San Jose scale gives adequate control. However, spray with methyl EC (0.25%), chlorpyriphos 20 EC (0.02%), dimethoate 30 EC (0.03%), malathion 50 EC (0.05%) in late May or early June on young trees.
Woolly Aphid: (*Eriosoma lanigerum*) Woolly aphids suck the sap of branches, twigs and roots. Galls are often formed at the point where aphids feed. Infested plants become sickly and stunted in growth.	i) In addition to dormant sprays for San Jose Scale, application of methyl demeton 25 EC (0.025%), diemthoate 30 EC (0.03%), malathion 50 EC (0.05%) between petal fall and June will prevent in festation during summer. Serious infestation near the harvest can be

	controlled by using dichlorvos 76 EC (0.05%) during summer. ii) Infestation on non bearing young trees (1-4 years old) can be reduced by placing carbofuran 3G or diamethoate 5G, 30g granules at 5 cm. Depth in the root zone in April-May when there is sufficient moisture in soil. In case no moisture is available give light irrigation.
Apple Leaf Miner: (*Gracilaria zachrysa*) Larvae mine the leaves. Tissue surrounding the mines becomes discoloured. If several mines are present on infested leaf may eventually shrivel, die and fall.	Spraying at green tip stage with spray oil against San Jose Scale will give effective control. Application of methyl demeton 25 EC (0.025%), dimethoate 30EC (0.03%) after petal fall will suppress the pest.
Hairy Caterpillar: (*Euproctis* spp.) Caterpillars feed leaves. In heavy infestation trees completely defoliated. Repeated defoliation may weaken and kill most of the trees.	i) Collection and destruction of egg masses. ii) Stappling burlap skirts around tree trunks infested with caterpillars and collection and destruction of larva/pupae from May to the end of June. iii) Spraying after petal fall chlorpyriphos 20 EC (0.02%) or quinalphos (0.02%)
Shot/ Pin Hole Borer: (*Scotytus amygda*) Adult and grubs tunnel into the sapwood and hardwood of the plant making galleries and pin holes. Surface of infested branches get perforated followed by yellowing and wilting of leaves. A serious infestation may kill the whole trees. Holes are some times indicated on the bark.	i) Pruning and destruction of borer infested branches. ii) Up-keep the vigor of trees by application of balanced dose of fertilizers will help in reducing the attack of shot hole borer/stem borer. iii) Leave scattered fallen tree trunks, and branches in the orchard for trapping egg laying scolytids during the month of June or first week of July. iv) Swabbing infested branches and trunks with 1.5% lindane dust and soil in the ration of 1:6 mixed with rice wheat straw.
Stem Borer: (*Apriona cinerea*) The beetles damage the stem branches drilling big holes in the trunk. Saw dust is seen coming out from holes.	To avoid serious damage by stem borer: Clean the hole and plug it with cotton soaked in petrol/dichlorovos after inserting naphthalene balls.Swabbing as in Shot/Pin hole borer.

Chaffer Beetles: (*Adoretus* spp.) Adult beetle feed on leaves, buds, blossom and fruit-lets. Eaten away leaves are perforated. The grubs feed on roots and may cause wilting of the plant.	i) Spray carbaryl 50WP (0.1%) during May-June at dusk. ii) Shaking of young non-bearing trees at dusk and collecting and destroying the beetles is useful. iii) Apply carbofuran 3G @ 1kg a. i/ha in soil against grubs.
Mites European Red Mite (*Panonychus ulmi*) **Two spotted spider mite:** (*Tetranychus urticae*) The mites puncture the tissues of the leaves and feed on plant juice. If the infestation is light the foliage becomes speckled heavily infested leaves become yellowish or bronzed. Severe infestation will result in early dropping of leaves, retarted growth, weakened fruit buds and reduced size of fruits.	Dormant oil emulsion spray or delayed dormant spraying with mixture of tree spray oil and organophosphorous compounds will provide adequate control of European red mite. Application of chlorpyriphos (0.02%), endosulfan 0.07% or quinalphos (0.025%) or dichlorvos (0.05%) in the last week of July or August applied against second brood of San Jose Scale may also provide effective control of mite pest complex. Spraying of dicofol (0.02%) may also prove effective.Certain fungicides like wettable sulphur, mancozeb also reduce the mite population.
Blossom Thrips: (*Thrips* spp.) Thrips feeding within the buds may cause weeping of sap from injured areas. Later brownish patches are produced at the bases of stamens and styles and on petal of infested blossom. Attacked leaf tissue also become distorted. Injury may lead to distortion of petals, restting of fruits and reduction of fruit set.	Control measures recommended for Sanjose Scale shall also take care of blossom thrips.
Pear Psylla: (*Psylla pyricola*) Heavy in festation leads to the development of small mishaped fruits and premature leaf fall or fruit drop. Blossom attacked by first generation nymphs turn later in the season may effect the following seasons pear crop by weakening or killing the development of fruit buds. The succulent tissues at the tips of new shoots are particularly liable to attack and tree growth may	Light infestation may be controlled by spraying about 3 weeks after petal fall with chloryriphos 20 EC (0.02%), methyl demeton 25 EC (0.025%), dimethoate 30 EC (0.03%) or malathion 50 EC (0.05%). Heavy infestation may require two sprays, one at petal fall and other three weeks later.

be checked. Sooty mould grows profusely on honey dew secreted by nymphs often coating the shoots, foliage and fruits in an unsightly and unpleasant black layer.

Economic Analysis

The information with respect to yearly costs involved in the establishment of pear orchards with spacing of 6 m × 6 m are as:

Cost of establishment

During the first year of the plantation, orchardists had incurred expenditure on the preparation of land, planting material, digging, filling of pits and planting, irrigation, training/ pruning, plant protection, farm yard manure, fertilizers, intercultural operations etc.The operation wise per hectare average costs for the first year is presented in table 35.The per hectare total costs incurred during first year for the establishment of pear orchards were found to be ₹ 39238.64. Variable costs per hectare were found to be ₹ 33078.64 with ₹ 5560.00 incurred on digging, filling and planting, ₹ 4000.00 on preparation of land, ₹ 11120.00 on planting material, ₹ 1200.00 on irrigation, ₹ 3336.00 on farm yard manure, ₹ 318.50 on fertilizers, ₹ 1000.00 on plant protection chemicals, and ₹ 1500.00 on training/pruning whereas fixed costs were equal to ₹ 6160.00 per hectare with ₹ 2250.00 incurred on depreciation of implements and ₹ 3250.00 as earned value of rented land (EVRL). The per hectare interest on working and fixed capital was worked out to be ₹ 3544.14 and ₹ 660.00, respectively.

Operational costs

The per hectare operational costs consists of yearly expenses on the maintenance of bearing orchard. The item wise and concept wise operational costs for pear orchards are presented in table 36. The per hectare variable costs and fixed were found to be ₹ 44022.81 and ₹ 5320.00, respectively with a total of ₹ 49342.81 per hectare. The table further revealed that among the working costs, the highest expenditures of ₹ 16680.00 per hectare were incurred on farm yard manure and lowest of ₹ 1500.00 on irrigation. The fixed costs included ₹ 500.00 on depreciation on machinery and farm inventory, ₹ 3250.00 as earned value of rented land and ₹ 570.00 as interest on fixed capital

Year wise cost and return analysis

The per hectare per year cost and returns from pear orchards worked out for different age groups are presented in Table 37. The returns starts from 4th year of the establishment of the orchard with average production of 13.90 quintals per hectare which increases to 111.20 quintals per hectare from 11th year and onwards. As far as the net returns are concerned, it was found to be ₹ 18503.52 per hectare in during 4th which increases to ₹ 395457.19 per hectare during 11th year of the establishment of the orchard and onwards.The benefit cost ratio of 1:1.50 was found during 4th which increases to 1:9.01 during 11th year of the establishment of the orchard which proves the viability of the orchard.

Table 35 Operation wise first year establishment costs of pear orchards with spacing = 6m x 6m (`/ha) No. of Plants per hectare = 278

S.No.	Item	Cost(in ₹)
1	Preparation of land	4000.00
2	Digging, filling and planting	5560.00
3	Planting material/ Plantation cost	11120.00
4	Irrigation	1200.00
5	Farm Yard Manure	3336.00
6	Fertilizers (Urea, DAP, MoP)	318.50
7	Plant Protection Chemicals	1000.00
8	Intercultural operations (weeding, hoeing etc)	1500.00
9	Training/ Pruning	1500.00
10	Interest on working capital @12 per cent per annum	3544.14
A	Total variable cost (1-10)	33078.64
11	Implements/ Depreciation	2250.00
12	Earned value of rented land	3250.00
13	Interest on fixed capital @ 12 per cent per annum	660.00
B	Total fixed cost (11-13)	6160.00
14	Total establishment cost (A+B)	39238.64

Table 36 Item wise and concept wise operational costs of pear orchards with spacing = 6m x 6m (`/ha)

S.No.	Item	Cost (in ₹)
1	Labour	4500.00
2	Intercultural operations and hoeing, weeding	2500.00
3	Irrigation	1500.00
4	Training/ pruning	2500.00
5	Farm Yard Manure	16680.00
6	Fertilizers (Urea, DAP, MoP)	9626.08
7	Plant protection	2000.00
8	Interest on working capital	4716.73
A	Total variable cost (1-8)	44022.81
9	Depreciation	1500.00
10	EVRL*	3250.00
11	Interest on fixed capital	570.00
B	Total fixed cost (9-11)	5320.00
12	Total operational/ maintenance cost (A+B)	49342.81

Table 37 Year wise cost and return analysis of pear orchards with spacing= 6m x 6m (` /ha)

S.No.	Particulars	Years							
		1	**2**	**3**	**4**	**5**	**6**	**Average 7th to 10th year**	**11th year and onwards**
1.	Total cost	39238.64	32646.84	34871.66	37096.48	39646.02	42593.20	48303.81	49342.81
2.	Production (qtls)	0.00	0.00	0.00	13.90	27.80	42.00	76.45	111.20
3.	Gross returns (in ₹)	0.00	0.00	0.00	55600.00	111200.00	168000.00	305800.00	444800.00
4.	Net returns (in ₹)	-39238.64	-32646.84	-34871.66	18503.52	71553.98	125406.80	257496.19	395457.19
5.	Benefit cost ratio	0.00	0.00	0.00	1:1.50	1:2.80	1:3.94	1:6.33	1:9.01

10

Cultivation, Management and Economic Analysis of Peach

Peach (*Prunus persica* (L.) Batsch.) is a highly popular fruit crop grown in temperate regions of the world. In India, peach cultivation is confined to mid hill zone of Himalayas extending from J&K State to Khasi Hill at an altitude of 1500 to 2000 meters above mean sea level. Low chilling cultivars are also grown in sub mountainous zones of Punjab, Haryana, Delhi and Western U.P. and also gaining popularity in sub-tropical areas of Jammu. Peach crop in J&K state covers an area of 2,615 hectares with 1,699 hectares falling in Jammu Division. Total production of peach in J&K state is 6,221 metric tonnes, out of which 2,612 metric tonnes are being contributed by Jammu Division. Main peach growing areas in Jammu region are hilly areas of Poonch, Doda,Kishtwar, Ramban Udhampur,Reasi, Rajouri and Kathua districts. Recently, the new orchards of low chill peaches are also being planted in Jammu, Samba and Kathua districts.

Climate and Soil

Peach does well in wet and cool climate with cold winter and dry summer. It likes mild climate.The limiting factors in peach cultivation are the minimum winter temperature, chilling hours, spring frost, hail storms, high humidity and desiccating winds during summer. In sub-tropical areas of North, the low chilling cultivars require 50-300 hours of chilling during the winter for proper flowering and fruiting.

The peach thrives well on light textured soil. Deep, fertile, loam or sandy loam with good drainage is considered well for successful peach cultivation. Poorly drained soil should be rejected for peach cultivation.

Cultivars

a) Sub-tropical areas

Early season (Maturing in last week of April): Florda Sun and Florda Prince

Mid season (Maturing in 1st to 2nd week of May): Shan-e-Punjab and Early Grand.

Late season (Maturing in mid June): Florda Red and Sharbati.

b) Temperate areas

Early Season (Maturing in third week of June): Saharanpuri (Saharanpuri No. 1) and World's Earliest(Saharanpuri No. 2)

Mid Season (Maturing in 1st to 2nd week of July): Peshawari and July Elberta (Amba-Chanun No.1)

Late Season (Maturing in last week of July to 1st week of August): Elberta (Amba Chanun No. 2) and Quetta

Rootstock

Rootstocks used for peach are wild peach, apricot, plum and peach. Plum stocks can be used when peaches are to be planted in a slightly heavier soil. Rootstock, in sub-tropical areas are generally grown from the seeds of desi peach tree or some commercial cultivars like Sharbati and Khurmani. Seed should be extracted from ripe fruit only, which ensures better germination. Peach seed needs stratification for germination which can be done by placing them under moist condition at about 1-5^{0}C for 60-70 days, however, this period varies with cultivars. Seeds can also be directly sown in the open ground in the month of November for stratification under natural conditions. Pre-soaking of peach seed in GA_3 at 100 ppm for 24 hours before placing them under stratification improves germination and reduces time for germination.

Propagation

Peach seeds become buddable by June to July when they attain pencil thickness. Peaches are propagated commercially by tongue (when stock and scion are of same size) or cleft grafting (when stock is thicker than scion) and T-budding or ring budding. In hills, tongue grafting during January-February and T-budding during June to September are performed. However, in plains grafting is performed during November-January and budding during April-June and September.

Planting

Planting should be done 6 metres apart in pits of 1m x 1m x 1m dug few months in advance of plantation. After planting, the young peach plants need staking with bamboo or wooden logs so that they can grow upright and straight.

The young plants should be watered at 15 days interval. If there is any indication of white ant attack, chlorpyriphos 1ml per litre of water with irrigation should be given.

After care

The peach is highly susceptible to scorching sun heat injury in sub-tropical areas, hence the open stem need to be whitewashed during summer. The problem of stem bark cracking can also be reduced by keeping plants low headed.

Training and Pruning

The peaches are trained to open centre or central modified leader system. For sub-tropical conditions where the sunshine is plenty and there are strong winds during summer, open centre system is not suitable in view of weak framework. Plants should be headed back to a height of 90-100 cm at the time of planting in the fields. All the branches in the plant are cut back to two buds. Within some weeks, the plants develop branches and 3-5 branches are retained well anchored around the trunk and selected as main branch. The lowest branch should be about 50 cm above from the ground level. In the second year, very little pruning is done and only overcrowded and interfering branches are removed. No lateral is allowed to outgrow the other.

In the third year, the main leader is suppressed by cutting it back to the nearest lateral.

Pruning in young and old trees

Peaches bear mostly on one year old growth. Pruning should be done annually so as to produce 50-100 cm of growth in young trees and 30-70 cm of growth in old trees. The heading back is accompanied by thinning out so that adequate sunlight enters into the tree. Pruning should be carried out in early January.

Fertilization

The peaches have high requirement for nitrogen and potassium as compared to phosphorous.

Age of tree (Year)	Fertilizer doses (g/tree)		
	Urea	Diammonium Phosphate	Muriate of Potash
1	55	20	65
2	115	40	130
3	170	65	200
4	225	85	265
5	280	110	330
6	395	150	585
7	510	195	750
8	620	240	915
9	735	280	1080

Note: Muriate of potash and diammonium phosphate should be applied as single dose in December and half dose of urea should be given 2-3 weeks before normal bloom, while rest half of urea be applied 3 weeks after fruit set.

Irrigation

Peach tree should be irrigated 2-3 times in a month during summer for obtaining optimum yield and quality fruit. The orchard should be provided with proper drainage system and water should not be allowed to stagnate.

In sub-tropical region, nine irrigations of 127 mm (40mm evapo-transpiration) applied from December to May improves the growth and quality of peach. After harvesting, the irrigation can be given at longer intervals.

Orchard floor management

The shallow cultivation of orchard will help in better soil aeration, destruction of weeds, increasing nitrification and destruction of hibernating insects. The tree should be white-washed regularly to protect from sunburn.

Harvesting and Yield

For harvesting peach crop the peculiar brown colour of the variety and 5% pit browning should be taken into consideration alongwith the experience of orchardist. For local market, the fruit should be picked at the full maturity, while for distant market the fruit should be harvested when slightly immature. All the peach fruit do not mature simultaneously and 3-4 pickings are required at 4-5 days interval. The yield may vary from 50 to 125 kg per tree depending upon the cultivars and orchard management practices. The peak harvesting period of peach in sub-tropical plains is from last week of April to mid of June.

Post harvest management

The peaches stored in polythene bags of 1006mm holes per m^2 improved the quality of fruit however, intermittent warming coupled with modified atmospheric storage is effective in preventing chilling injury and prolonging storage life of peach.

Diseases and their Management

Symptoms	Management
Leaf Curl: (*Taphrina deformans*) Symptoms develop on leaves, twigs,flowers and fruits.The leaf blade becomes thickened and puckered along with midrib leaf curls and turn syellowish. Later, the upper leaf surface (200g/100L water), chlorthalonil gives a silvery appearance and becomes brittle. Finally, the affected leaves dry and drop.Infected twigs become swollen and may exude a gum like substance. The affedted twigs usually die. Flowers and fruits, when infected, drop prematurely.	Spray the trees at leaf fall, bud swell, petal fall and fruit-let stage with: carbendazim (100g/ 100L water), captafol (200g/ 100L water), thiophanate methyl (100g/1001 water) or mancozeb (200g/100L water).
Coryneum Blight and Shoot Hole: (*Stigmina carpophyla*) Twigs, blossoms, leaves,fruits and unopened buds are blighted. Small, circular, deep purple spots appear on the fruit.The spots are raised and rough. On leaves, dark brown, scattered lesions enlarge rapidly.Abscission of the diseased areas of the leaves results in shot holes.The diseased buds become darker in colour. Small purplish, raised spotsdevelop on twigs which expand into necrotic cankers.	Spray the leaves at leaf fall, before bud-burst, fruitset and 15-20 days after fruitset stage with: copper oxychloride (300g/ 100L water), carbendazim (100g/100L water) or captafol (200g/100L water).

Insect Pest Management

Symptoms	Management
Peach Leaf Curl Aphid: (*Myzus persicae*) Nymphs and adults suck the sap from the leaves, petiole, blossom and fruits causing leaf curl and distortion. Blossoms wither and fruit do not develop and drop if formed. **Peach Tree Borer:** Larvae feed in irregular tunnels in the bark and outer wood of the trunk and main root near the ground surface. The injured area produces a great deal of gum usually containing casting from the borer. **Peach Twig Borer:** Larvas tunnel the buds and terminal shoots in early summer. The injured trees wilt and die. Larvae also infect the fruits from the stem and may bear into the fruit. **Plum Aphid:** The aphids suck the sap and weaken the trees, some species cause marked curling of leaves. Sooty fungus grows on honey dew exuded by aphids and thereby reducing the market value of fruit. **Green Peach aphid:** Aphids feed on leaves, blossom and fruitlets resulting in fruit drop and some times leaf curling. **Lecanium Scale:** The insect pierce the tissue and suck the sap and if abundant for 2-3 years, the infested branches may get killed. Mould develops on the exude of this pest. **Sanjose Scale:** All parts above ground are attacked heavily infested trees lack vigour its foliage turn thin and yellow speckled. Often bark around scale is reddened. On fruits the insect appear as grey patches surrounded by inflamed red area.	1.Keep the trees as free as possible from mechanical wounds, winter injury, crotch separation and cankers. 2.Destroy dead and dying fruit trees and branches to avoid borer infestation. 3.Prune and destroy the scale and borer infested branches. 4.Maintain vigour of the tree to keep away shoot hole/pin hole borer, bark beetles and other pest infestation. 5.Spray 6.33 diesel oil emulsion or 3% tree spray oil during dormancy when the temperature is above 4^0C to control scales, aphids and mites etc. 6.Spray at bud burst stage any of the insecticides, if heavy infestation of mites, aphids, scales etc. is observed with Chlorpyriphos (0.02%), methyldemeton (0.025%) and dimethoate (0.03%) 7.Spray 7-10 days after petal fall any of the insecticides if curling of leaves is noticed, chlorpyriphos (0.02%), dimethoate (0.04%), malathion (0.05%), methydemeton 25EC (0.025%) to control scales aphids, mites twig borer and other pests. Repeat after 2-3 weeks, if necessary. Insecticides applied at bud burst stage should not be repeated. 8.Fruit fly can be controlled by spraying the tree with dimethoate (0.03%), chlorpyriphos (0.02%) or malathion (0.05%) about 10 days after petal fall. Repeat with dichlorvos (0.05%) if necessary after sweet varieties begin to develop solour. 9.Swabing infested branches and trunk with 1.5% lindane dust and soil in the

Flat Headed Borer: Adults feed on foliage. Young grubs feed below the bark which gets loosened and splits. Full grown grubs bore into wood. Gum oozes out from the holes. Leaves pale and growth of the tree is arrested.

Short Hole Borer/Pin Hole Borer: Adults and grubs tunnel into the sapwood and hardwood of the plant making galleries and pin holes. Surface of infested branches get perforated followed by yellowing and wilting of leaves. A serious infestation may kill the whole tree. Holes are some times indicated by boring on the bark.

Hairy Caterpillar: Caterpillars feed on leaves. In heavy infestation trees may be completely defoliated. Repeat defoliation may weaken and kill most of the trees.

Chaffer beetles: Adult beetle feeds on leaves, buds, blossom and fruitlets. Eaten away leaves are perforated. The grubs feed on roots and may cause wilting of the plant.

Stem Borer: The beetles damage the stem branches drilling big holes in the trunk. Saw dust is seen coming out from holes.

Apricot Chalcid: The grubs bore into the kernels and feed in the inner contents leaving the papery coat in tact. As a result fruit development is arrested and the fruit fall prematurely with grubs still feeding within the fruits.

ration of 1:6 mixed with rice or wheat straw will help inreducing damage by shoot hole borer and other insect borers.

10. Apply chlorpyriphos 1.5% or lindane 1.5% dust at 20-25 kg/hect in the soil against the grubs of chaffer beetles.

11. To avoid serious damage of stem borers, clean the holes and plug them with cotton soaked in petrol after inserting naphthalene balls in the holes. Swab the affected area with 1.5% Lindane dust and soil in the ratio of 1:6.

12. Pre-bloom spraying against aphids and mites will also provide control against the apricot chalcid. However, all fallen fruits which may contain the grub be collected and burnt. Pre-bloom spraying against aphids may also provide some degree of control against peach twig borer. However, additional application of any insecticide listed at 6 may provide satisfactory control in case of heavy infestation.

13. Spray 10 days after petal fall and three weeks later methyl demeton (0.02%), dimethoate (0.03%) or chlorpyriphos (0.02%) against peach tree borer.

Economic analysis

The information with respect to yearly costs involved in the establishment of peach orchards with spacing of 6 m × 6 m is as:

Cost of establishment

During the first year of the plantation, orchardists had incurred expenditure on the preparation of land, planting material, digging, filling of pits and planting, irrigation, training/ pruning, plant protection, farm yard manure, fertilizers, intercultural operations etc. The operation wise per hectare average costs for the first year is presented in table 38. The per hectare total costs incurred during first year for the establishment of peach orchards were found to be ₹ 48008.64. Variable costs per hectare were found to be ₹ 39608.64 with ₹ 5560.00 incurred on digging, filling and planting, ₹ 6950.00 on preparation of land, ₹ 11120.00 on planting material, ₹ 1800.00 on irrigation, ₹ 3336.00 on farm yard manure, ₹ 419.36 on fertilizers, ₹ 2289.50 on plant protection chemicals, and ₹ 1500.00 on training/pruning whereas fixed costs were equal to ₹ 8400.00 per hectare with ₹ 2500.00 incurred on depreciation of implements and ₹ 5000.00 as earned value of rented land (EVRL). The per hectare interest on working and fixed capital was worked out to be ₹ 4243.78 and ₹ 900.00, respectively.

Operational costs

The per hectare operational costs consists of yearly expenses on the maintenance of bearing orchard. The item wise and concept wise operational costs for peach orchards are presented in table 39. The per hectare variable costs and fixed were found to be ₹ 40857.76 and ₹ 8400.00, respectively with a total of ₹ 49257.76 per hectare. The table further revealed that among the working costs, the highest expenditures of ₹ 16680.00 per hectare were incurred on farm yard manure and lowest of ₹ 1800.00 on irrigation.The fixed costs included ₹ 2500.00 on depreciation on machinery and farm inventory, ₹ 50000.00 as earned value of rented land and ₹ 900.00 as interest on fixed capital

Year wise cost and return analysis

The per hectare per year cost and returns from peach orchards worked out for different age groups are presented in Table 40. The returns starts from 4th year of the establishment of the orchard with average production of 19.46 quintals per hectare which increases to 111.20 quintals per hectare from 10th year and onwards. As far as the net returns are concerned, it was found to be ₹ 44949.51 per hectare during 4th which increases to ₹ 506742.24 per hectare during 10th year of the establishment of the orchard and onwards. The benefit cost ratio of 1:2.37 was found during 4th which increases to 1:11.29 during 10th year of the establishment of the orchard which proves the viability of the orchard.

Table 38 Operation wise first year establishment costs of peach orchards with spacing = 6m x 6m (₹/ha) No. of Plants per hectare = 278

S.No.	Item	Cost (in ₹)
1	Preparation of land	6950.00
2	Digging, filling and planting	5560.00
3	Planting material/ Plantation cost	11120.00
4	Irrigation	1800.00
5	Farm Yard Manure	3336.00
6	Fertilizers (Urea, DAP, MoP)	419.36
7	Plant Protection Chemicals	2289.50
8	Intercultural operations (weeding, hoeing etc)	2390.00
9	Training/ Pruning	1500.00
10	Interest on working capital @12 per cent per annum	4243.78
A	Total variable cost (1-10)	39608.64
11	Implements/ Depreciation	2500.00
12	Earned value of rented land	5000.00
13	Interest on fixed capital @ 12 per cent per annum	900.00
B	Total fixed cost (11-13)	8400.00
14	Total establishment cost (A+B)	48008.64

Table 39 Item wise and concept wise operational costs of peach orchards with spacing = 6m x 6m (₹/ha)

S.No.	Item	Cost (in ₹)
1	Labour	3900.00
2	Intercultural operations and hoeing, weeding	2500.00
3	Irrigation	1800.00
4	Training/ pruning	2500.00
5	Farm Yard Manure	16680.00
6	Fertilizers (Urea, DAP, MoP)	6600.14
7	Plant protection	2500.00
8	Interest on working capital	4377.62
A	Total variable cost (1-8)	40857.76
9	Depreciation	2500.00
10	EVRL*	5000.00
11	Interest on fixed capital	900.00
B	Total fixed cost (9-11)	8400.00
12	Total operational/ maintenance cost (A+B)	49257.76

* Earned value on rented land

Table 40 Year wise cost and return analysis of peach orchards with spacing= 6m x 6m (₹/ha)

S.No.	Particulars	Years							
		1	**2**	**3**	**4**	**5**	**6**	**Average 7th-9th year**	**10th year and onwards**
1.	Total cost	48008.64	29736.73	32109.45	32890.49	36794.17	40256.66	46550.75	49257.76
2.	Production (qtls)	0.00	0.00	0.00	19.46	41.70	69.50	111.20	111.20
3.	Gross returns (in ₹)	0.00	0.00	0.00	77840.00	168000.00	278000.00	444800.00	556000.00
4.	Net returns (in ₹)	-48008.64	-29736.73	-32109.45	44949.51	131205.83	237743.34	398249.25	506742.24
5.	Benefit cost ratio	0.00	0.00	0.00	1:2.37	1:4.57	1:6.91	1:9.56	1:11.29

11

Cultivation, Management and Economic Analysis of Plum

Plum (*Prunus saliciana* Lindl.) is a temperate fruit and grows on commercial scale in J&K, HP and Uttaranchal. In J&K, plum cultivation is spread over 4,038 hectares with total yield of 10,112 metric tones. In Jammu Division, plum is being grown in temperate areas of Poonch, Rajouri, Doda and Kathua districts covering an area of 2,665 hectares with an annual production of 14,240 metric tonnes.

Climate and Soil

Plums because of their wide variability between different species, exhibit considerate varietal differences in their climatic requirements.The Japanese plums do very well in the low and mid hill regions ranging between 650 to 1650 meters above mean sea level. Some varieties can be grown successfully in the warmer regions of plains.

For good growth and longer life of plants, well drained medium to deep soils are most suitable. Plums propagated on peach root stocks grow better on sandy loam soil, while for heavier soils apricot rootstock or Kabul green gage root stocks of plum can be used.

Cultivars

a) Sub-tropical areas

Titron, Kala Amritsari, Kataruchak and Alubukhara

b) Temperate areas

Sharp's Early, Formosa	(Early season)
Wickson, Satsuma, Burbank, Santa Rosa	(Mid season)
Blue Impeatrice, Prune-D-Agen, Grand Duke	(Late season)

Propagation and Rootstock

Wild plum stones can be used for rootstock, where a desired cultivar is T-budded on it in July-September. The seeds of plum need stratification for germination.

Planting should be done in dormant season, when danger of winter injury is over. Planting distance of 6 m x 6 m is recommended in a square system of layout. Pits 1 m x 1 m x 1 m are dug about a month before planting and are filled with fertile top soil mixed with 40 kg of well rotten farmyard manure per pit.

Training and Pruning

The tree should be trained to modified leader system. The plant is headed back to the height of 90 cm from ground level. After planting in the field, two years old tree have a number of laterals, and 4 or 5 of which are selected to make a proper framc work. Pruning in 3rd and 4th year is done to keep a balance between main scaffold limbs and secondary branches. During 5th year, leader should be headed back at a point where a suitable out growing lateral is present to check the height of the plants.

Pollination and fruit setting

The Japanese plum cultivars require cross-pollination to set satisfactory crop.However, Santa Rosa, and Kala Amritsari, are self-fruitful grown in temperate and sub-tropical areas, respectively. But in both varieties yield can be improved if cross compatible variety is also planted in the orchard.

Fertilization

Age of tree (Year)	Fertilizer doses (g/tree)		
	Urea	Diammonium Phosphate	Muriate of Potash
1	55	20	65
2	115	40	130
3	170	65	200
4	225	85	265
5	280	110	330
6	395	150	585
7	510	195	750
8	620	240	915
9	735	280	1080

Note: Muriate of potash and diammonium phosphate should be applied as single dose in December and half dose of urea should be given 2-3 weeks before normal bloom, while rest half of urea be applied 3 weeks after fruit set.

Irrigation

The irrigation should be avoided during flowering and fruit-setting stage. During hot months, frequent irrigation should be given for the development of fruit size and quality. Irrigation should be stopped at colour break stage to avoid heavy fruit drop. Orchard should have proper drainage system.

Harvesting and Yield

For harvesting plum crop, following maturity guide should be adopted:-

1. *Ground colour:* Ground colour changes with maturity depending upon varieties.
2. *Pit browning:* In case of free stone cultivars, pit-browning upto 5% of the pit area is a good index of maturity.

Plum starts yield within 3 years of planting and in general fruit ripe stage is reached 12 weeks after fruit set. In sub-tropical areas, peak harvesting time is 2nd fortnight of May. In temperate zone, the fruit ripen within 92 to 136 days after fruit set depending upon the cultivars and average daily temperature. The grown up tree of plum yields about 30-50 kg of fruit per tree.

Effective preservation of fresh fruits in condition is mainly governed by harvesting of fruits at right stage of maturity and post-harvest treatments.

Diseases and their Management

Symptoms	Management
Leaf Curl: (*Taphrina deformans*) Symptoms develop on leaves, twigs,flowers and fruits. The leaf blade becomes thickened and puckered along with midrib leaf curls and turn syellowish.Later, the upper leaf surface (200g/100L water), chlorthalonil gives a silvery appearance and becomes brittle. Finally, the affected leaves dry and drop. Infected twigs become swollen and may exude a gum like substance. The affedted twigs usually die. Flowers and fruits, when infected, drop prematurely.	Spray the trees at leaf fall, bud swell, petal fall and fruit-let stage with: carbendazim (100g/ 100L water), captafol (200g/ 100L water), thiophanate methyl (100g/1001 water) or mancozeb (200g/100L water).
Coryneum Blight and Shoot Hole: (*Stigmina carpophyla*) Twigs, blossoms, leaves,fruits and unopened buds are blighted. Small, circular, deep purple spots appear on the fruit. The spots are raised and rough.	Spray the leaves at leaf fall, before bud-burst, fruitset and 15-20 days after fruitset stage with: copper oxychloride (300g/

On leaves, dark brown, scattered lesions enlarge rapidly. Abscission of the diseased areas of the leaves results in shot holes.The diseased buds become darker in colour. Small purplish, raised spotsdevelop on twigs which expand into necrotic cankers.	100L water), carbendazim (200g/100L water). (100g/100L water) or captafol

Insect Pest Management

Symptoms	Management
Peach Leaf Curl Aphid: Nymphs and adults suck the sap from the leaves, petiole, blossom and fruits causing leaf curl and distortion. Blossoms wither and fruit do not develop and drop if formed. **Peach Tree Borer:** Larvae feed in irregular tunnels in the bark and outer wood of the trunk and main root near the ground surface. The injured area produces a great deal of gum usually containing casting from the borer. **Peach Twig Borer:** Larvas tunnel the buds and terminal shoots in early summer. The injured trees wilt and die. Larvae also infect the fruits from the stem and may bore into the fruit. **Plum Aphid:** The aphids suck the sap and weaken the trees, some species cause marked curling of leaves. Sooty fungus grows on honey dew exuded by aphids and thereby reducing the market value of fruit. **Green Peach Aphid:** Aphids feed on leaves, blossom and fruitlets resulting in fruit drop and some times leaf curling. **Lecanium Scale:** The insect pierce the tissue and suck the sap and if abundant for 2-3 years, the infested branches may get killed. Mould develops on the exude of this pest. **Sanjose Scale:** All parts above ground are attacked heavily infested trees lack vigour its foliage turn thin and yellow	1.Keep the trees as free as possible from mechanical wounds, winter injury, crotch separation and cankers. 2.Destroy dead and dying fruit trees and branches to avoid borer infestation. 3.Prune and destroy the scale and borer infested branches. 4.Maintain vigour of the tree to keep away shot hole/pin hole borer, bark beetles and other pest infestation. 5.Spray 6.33% diesel oil emulsion or 3% tree spray oil during dormancy when the temperature is above 4^0C to control scales, aphids and mites etc. 6.Spray at bud burst stage any of the insecticides, if heavy infestation of mites, aphids, scales etc. is observed, with chlorpyriphos 20 EC (0.02%), methyl demeton 25 EC (0.025%) and dimethoate 30 EC (0.03%). 7.Spray 7-10 days after petal fall any of the insecticides if curling of leaves is noticed, chlorpyriphos (0.02%), dimethoate (0.04%), formothion (0.025%), malathion (0.05%), endosulfan 35 EC (0.07%), methyl demeton (0.025%) to control scales aphids, mites twig borer and other pests.Repeat after 2-3 weeks, if necessary. Insecticides applied at bud burst stage should not be repeated. 8.Fruit fly can be controlled by spraying the tree with endosulfan (0.07%) or

speckled. Often bark around scale is reddened. On fruits the insect appear as grey patches surrounded by inflamed red area.

Flat Headed Borer: Adults feed on foliage. Young grubs feed below the bark which gets loosened and splits. Full grown grubs bore into wood. Gum oozes out from the holes. Leaves pale and growth of the tree is arrested.

Shoot Hole Borer/Pin Hole Borer: Adults and grubs tunnel into the sapwood and hardwood of the plant making galleries and pin holes. Surface of infested branches get perforated followed by yellowing and wilting of leaves. A serious infestation may kill the whole tree. Holes are some times indicated by boring on the bark.

Hairy Caterpillar: Caterpillars feed on leaves. In heavy infestation trees may be completely defoliated. Repeat defoliation may weaken and kill most of the trees.

Chaffer Beetles: Adult beetle feeds on leaves, buds, blossom and fruitlets. Eaten away leaves are perforated. The grubs feed on roots and may cause wilting of the plant.

Stem Borer: The beetles damage the stem branches drilling big holes in the trunk. Saw dust is seen coming out from holes.

Apricot Chalcid: The grubs bore into the kernels and feed in the inner contents leaving the papery coat in tact. As a result fruit development is arrested and the fruit fall prematurely with grubs still feeding within the fruits.

dimethoate (0.03%), chlorpyriphos (0.02%) or malathion (0.05%) about 10 days after petal fall. Repeat with dichlorvos (0.05%) if necessary after sweet varieties begin to develop colour.

9.Swabing infested branches and trunk with 1.5% lindane dust and soil in the ratio of 1:6 mixed with rice or wheat straw will help in reducing damage by shot hole borer and other insect borers.

10.Apply 1.5% lindane dust at 20-25 kg/ha or carbofuran 3G at 1.5 kg a.i/ha in the soil against the grubs of chaffer beetles.

11.To avoid serious damage of stem borers, clean the holes and plug them with cotton soaked in petrol after inserting naphthelen balls in the holes. Swab the affected area with 1.5% Lindane dust and soil in the ratio of 1:6.

12.Pre-bloom spraying against aphids and mites will also provide control against the apricot chalcid. However, all fallen fruits which may contain the grub be collected and burnt. Pre-bloom spraying against aphids may also provide some degree of control against peach twig borer. However, additional application of any insecticide listed at 6 may provide satisfactory control in case of heavy infestation.

13.Spray 10 days after petal fall and three weeks later methyl demeton (0.025%), dimethoate (0.03%) or chlopyriphos (0.02%) against peach tree borer.

Economic Analysis

The information with respect to yearly costs involved in the establishment of plum orchards with spacing of 6 m × 6 m is as:

Cost of establishment

During the first year of the plantation, orchardists had incurred expenditure on the preparation of land, planting material, digging, filling of pits and planting, irrigation, training, plant protection, farm yard manure, fertilizers, intercultural operations etc. The operation wise per hectare average costs for the first year is presented in table 41.The per hectare total costs incurred during first year for the establishment of plum orchards were found to be ₹ 43016.24. Variable costs per hectare were found to be ₹ 35176.24 with ₹ 5000.00 incurred on digging, filling and planting, ₹ 6950.00 on preparation of land, ₹ 11120.00 on planting material, ₹ 1500.00 on irrigation, ₹ 1668.00 on farm yard manure, ₹ 419.36 on fertilizers, ₹ 2250.00 on plant protection chemicals, and ₹ 1500.00 on training whereas fixed costs were equal to ₹ 7840.00 per hectare with ₹ 2000.00 incurred on depreciation of implements and ₹ 5000.00 as earned value of rented land (EVRL). The per hectare interest on working and fixed capital was worked out to be ₹ 3768.88 and ₹ 840.00, respectively.

Operational costs

The per hectare operational costs consists of yearly expenses on the maintenance of bearing orchard. The item wise and concept wise operational costs for plum orchards are presented in table 42. The per hectare variable costs and fixed were found to be ₹ 34443.44 and ₹ 7840.00, respectively with a total of ₹ 42273.44 per hectare. The table further revealed that among the working costs, the highest expenditures of ₹ 13344.00 per hectare were incurred on farm yard manure and lowest of ₹ 1500.00 on intercultural operations. The fixed costs included ₹ 2000.00 on depreciation on machinery and farm inventory, ₹ 5000.00 as earned value of rented land and ₹ 840.00 as interest on fixed capital

Year wise cost and return analysis

The per hectare per year cost and returns from plum orchards worked out for different age groups are presented in Table 43. The returns starts from 4th year of the establishment of the orchard with average production of 6.95 quintals per hectare which increases to 97.30 quintals per hectare from 10th year and onwards. As far as the net returns are concerned, it was found to be ₹ 80728.38 per hectare during 5th year which increases to ₹ 346926.56 per hectare during 10th year of the establishment of the orchard and onwards. The benefit cost ratio of 1:3.65 was found during 5th year which increases to 1:9.21 during 10th year of the establishment of the orchard which proves the viability of the orchard.

Table 41 Operation wise first year establishment costs of plum orchards with spacing = 6m x 6m (₹/ha) No. of Plants per hectare = 278

S.No.	Item	Cost (in ₹)
1	Preparation of land	6950.00
2	Digging, filling and planting	5000.00
3	Planting material/ Plantation cost	11120.00
4	Irrigation	1500.00
5	Farm Yard Manure	1668.00
6	Fertilizers (Urea, DAP, MoP)	419.36
7	Plant Protection Chemicals	2250.00
8	Intercultural operations (weeding, hoeing etc)	1000.00
9	Training	1500.00
10	Interest on working capital @12 per cent per annum	3768.88
A	Total variable cost (1-10)	35176.24
11	Implements/ Depreciation	2000.00
12	Earned value of rented land	5000.00
13	Interest on fixed capital @ 12 per cent per annum	840.00
B	Total fixed cost (11-13)	7840.00
14	Total establishment cost (A+B)	43016.24

Table 42 Item wise and concept wise operational costs of plum orchards with spacing = 6m x 6m (₹/ha)

S.No.	Item	Cost (in ₹)
1	Labour	3500.00
2	Intercultural operations and hoeing, weeding	1500.00
3	Irrigation	1800.00
4	Training/ pruning	2000.00
5	Farm Yard Manure	13344.00
6	Fertilizers (Urea, DAP, MoP)	6600.14
7	Plant protection	2000.00
8	Interest on working capital	3689.28
A	Total variable cost (1-8)	34433.44
9	Depreciation	2000.00
10	EVRL*	5000.00
11	Interest on fixed capital	840.00
B	Total fixed cost (9-11)	7840.00
12	Total operational/ maintenance cost (A+B)	42273.44

* Earned value on rented land

Table 43 Year wise cost and return analysis of plum orchards with spacing= 6m x 6m (₹/ha)

S.No.	Particulars	Years							
		1	2	3	4	5	6	Average 7th to 9th year	10th year and onwards
1.	Total cost	41504.24	23435.69	25775.68	28114.74	30471.62	33964.50	39968.22	42273.44
2.	Production (qtls)	0.00	0.00	0.00	6.95	27.80	41.70	69.50	97.30
3.	Gross returns (in ₹)	0.00	0.00	0.00	27800.00	111200.00	166800.00	278000.00	389200.00
4.	Net returns (in ₹)	-41504.24	-23435.69	-25775.68	-314.74	80728.38	132835.50	238031.78	346926.56
5.	Benefit cost ratio	0.00	0.00	0.00	1:0.99	1:3.65	1:4.91	1:6.96	1:9.21

❑❑❑

12

Cultivation, Management and Economic Analysis of Walnut

Walnuts (*Juglans regia* L.) are found in the northern parts of the country in temperate belts. They are grown in the length and breadth of the valley in Jammu and Kashmir and hilly areas of Jammu. The walnut is being cultivated in the state on 89,339 ha with an annual production of about 2,66,280 metric tonnes. In Jammu Division, walnut plantation is on 38,318 ha with an annual production of 92,227 metric tonnes. The walnut is very rich in proteins, fats and minerals and is a concerned source of energy.

Climate and Soil

Walnuts are found growing in all parts of the Himalayan region between the elevation of 1200 to 2100 m above mean sea level. Walnuts are sensitive to both low and high temperature.Climate should be free from frost in spring and from extreme heat in summer. They grow well in areas where rainfall is about 75 cm or more and is well distributed. Temperature of 29-32°C near harvest time, results in well-filled kernels.

Walnuts do best in soils where roots can penetrate deep evenly. In addition to good top soil, sub-soil must be free of rock, impervious clay or layers of gravel. Most suitable soil is well drained silt loam with abundant organic matter. Alkaline soil and sandy soils should be avoided for walnut growing. A soil with 2-3 metres depth and pH range of 6-7 gives best results.

Cultivars

Most of the trees are of seedling origin and no standard cultivar is grown at present. Usually walnuts are grouped into 2 broad categories.

i) Thin shelled (Burzal)

ii) Medium shelled (Kagzi)

The following introductions have been found promising

i) Lake English

ii) Jug drajnoviski

Propagation

Walnuts can be propagated either by seed or vegetative methods.

The walnuts should be collected from vigorous and high yielding trees. The selected walnuts should be of big size, bright brown colour and good cracking quality of the shell, bright yellow colour, good taste and flavour of the kernel. The nuts are stratified by placing in alternate layers with suitable medium and subjected to cold treatment 3-4^{0}C for a period of 2-3 months.

Walnuts are stratified till December and are then sown immediately at 8-10 cm depth in well prepared beds of 30 cm in row to row and 10 cm seed to seed. The germination starts in the beginning of March. The planting of large nuts at 7 cm depth in soil in vertical position gives better germination and seedling growth.

Vegetative propagation

Walnut can be propagated vegetatively by grafting, budding and tooling. *Juglans regia* is the most easily available rootstock for walnut in India and is the only one used whenever walnut trees are raised by budding or grafting. This rootstock has better compatibililty and is practically suitable for light soils however it is unsuitable for waterlogged, saline and nematode infected soils. The most common method of grafting practiced in this nut is cleft grafting. Better survival of grafts is obtained when grafting operation is carried out in the middle of January. Patch budding is common among the budding methods which gives 30-35% success. Mid June is found to be an optimum time for budding.

Planting

Walnuts should be planted by square or hexagonal systems in plains and contour system is recommended for undulating areas. Pits of size 1 m x 1m x 1m are dug and then filled with well rotten FYM and soil in 1:2 proportion. Planting is done in early spring. Budded/grafted plants should be planted 10.5 m apart.

Training and Pruning

Walnut trees are trained according to modified leader system. Pruning of walnut tree is determined by the bearing habit. In the terminal bearing cultivars, pruning consists of heading back of selected framework branches and thinning out of competing limbs. The lateral bearing cultivars are both thinned and headed back heavily in order to encourage shoot growth which get suppressed due to early fruit production. The pruning operations are carried out in the dormant season.

Irrigation

Watering is very essential for the establishment of grafts and young plants. The water requirement, however, decreases with the development of roots. The frequency of irrigation depends upon the age of the tree, type of soil, rainfall, mulching and soil management system. The choice of an irrigation system depends upon water supply, ground level, and above all cost benefit ratio.

Flowering and fruit sets

Walnut seedling take about eight to ten years to come into bearing, but grafted plants begin fruiting early in about four to five years.

Fertilization

For achieving higher yields of quality crop, following recommendation of fertilizers should be applied to walnut trees.

Age of tree (Year)	Fertilizer doses (g/tree)		
	Urea	Diammonium phosphate	Muriate of Potash
1-3	20-40	60-65	30-50
4-6	60-135	65	65-125
7-10	190-490	65-130	165-415
11-15	585-1565	160-435	500-995
16-19	1960-3415	540-975	1245-2155
15th year and above	3915	1085	2490

Note: Full dose of DAP and MOP and ½ dose of urea should be applied two weeks before expected bloom. Rest half quantity of urea should be split into two equal doses which should be applied 3 weeks after fruit set and during early July.

Harvesting and Yield

Walnut is mostly harvested in the months of September and October. Some nuts fall on the ground after splitting of the hulling a natural way, while others are forced to drop down by beating with long poles. The total harvest is done in two or three times at few days interval. Pre-harvest treatment of walnut trees with ethrel @ 200-300 ppm at packing tissue turned brown (PTB) stage proved better to induce early and uniform hull dehiscence (dehulling) in walnuts. Moreover, it makes harvest of entire crop possible at one time and advances dehulling by about 12 days. The average yield of walnut is 40 kg per tree but a larger tree of about 40 year age can produce as much as 175 kg nuts per tree.

Diseases and their Management

Symptoms	Management
Leaf Blotch: (*Marssonina juglans*) The affected leaves show small, circular, light brown spots which later develop into large blotches having irregular outline and often involve the entire lamina.The pathogen survives mainly on the infected portions of dead,fallen and over wintered leaves.	i) Collect and burn fallen leaves in autumn. ii) Spray the tree with thiophanate methyl, carbendazim (50g/100L water) or dodine (75g/100L water) in late spring. Repeat 2-3 sprays at 15-20 days interval.
Mistletoe: (*Visum album*) The phanerogamic parasite forms witches broom on the tree and feeds on walnut plant reducing its yield.	The parasitic plant should be removed from the tree along with haustoria. In case of severe infestation, the entire branch or twig should be removed and burnt.

Insect Pest Management

Symptoms	Management
Mealy Bugs: Presence of insects covered with white cottony mass below the collar of the plant. Wilting and drying of nursery plants. Old and weak plantations are more susceptible to attack and exhibit nutrient deficiency symptoms and sickly appearance which result in poor yield. **Leaf Curl Aphid:** Nymphs and adults suck the sap from the leaves, petiole, blossom and fruits causing leaf curl and distortion. Blossoms wither and fruit do not develop and drop, if formed. **Green Peach Aphid:** Aphids feed on leaves, blossom and fruitlets resulting in fruit drop and some times leaf curling. **Lecanium Scale:** The insects pierce the tissue and suck the sap and if abundant for 2-3 years, the infested branches may get killed. Mould develops on the exude of this pest. **Sanjose Scale:** All parts above ground are attacked, heavily infested trees lack vigour its foliage turn thin and yellow speckled. Often bark around scale is reddened. On fruits the insect appear as grey patches surrounded by inflamed red area. **Hairy Caterpillar:** Caterpillars feed on leaves. In heavy infestation trees may be completely defoliated. Repeated defoliation may weaken and kill most of the trees. **Chaffer Beetles:** Adult beetles feed on leaves, buds, blossoms and fruitlets. Eaten away leaves are perforated. The grubs feed on roots and may cause wilting of plant.	1. Avoid planting of insect infested saplings. 2. Remove dead drying trees. 3. Pruning and destruction of borer infested branches. 4. Collection & destruction of egg masses of hairy caterpillar. 5. Keep up vigour of trees by application of balanced doses of fertilizers, this will help to reduce the attack of borers & other pests. 6. Stapling burlap skirts around tree trunks infected with hairy caterpillars and collection of larvae pupae from May to end of June and their destruction will help in suppressing the pest population. 7. Wherever possible spray the trees with chlorpyriphos (0.02%), or endosulfan (0.07%) in the month of May to reduce the population of aphids, mites, hairy caterpillars, walnut weevil and other pests. Repeat the spray after 3-8 weeks, if required. 8. Collect and destroy the fallen walnut leaves. The leaf size is reduced and turns yellowish which fall prematurely.

Stem Borer: The beetles damage the stem, branches by drilling big holes in the trunk. Saw dust is seen coming out from holes.

Shot Hole /Pin Hole Borer: Adults & grubs tunnel into the sapwood and hardwood of the plant making galleries and pin holes. Surface of infested branches get perforated followed by yellowing and wilting of leaves. A serious infestation may kill the whole tree. Holes are some times indicated by boring on the bark.

Walnut Aphids: The insects suck the sap from leaves which results in drying, wilting and yellowing of foliage. The infested leaves fall prematurely.

Grey Weevil: The weevil nibbles walnut leaves. Nursery plants are more affected; continued feeding may defoliate a nursery plant.

Walnut weevils: (*Alcidodus porrectirostis*) Eggs are laid in the pits excavated on the fruits. The grubs bore into the fruits and feed on kernel. The adult bite circular hole.

Collection and destruction of fallen nuts during May-June.Proper sanitation should be maintained.Smaller and medium plants can be sprayed with 0.02% chlorpyriphos 20EC or 0.05% dichlorvos during the month of April-May.

Economic Analysis

The information with respect to yearly costs involved in the establishment of walnut orchards with spacing of 10 m × 10 m is as:

Cost of establishment

During the first year of the plantation, orchardists had incurred expenditure on the preparation of land, planting material, digging, filling of pits and planting, irrigation, training, plant protection, farm yard manure, fertilizers, intercultural operations etc.The operation wise per hectare average costs for the first year is presented in table 44.The per hectare total costs incurred during first year for the establishment of walnut orchards were found to be ₹ 37768.08. Variable costs per hectare were found to be ₹ 31048.08 with ₹ 6000.00 incurred on digging, filling and planting, ₹ 5000.00 on preparation of land, ₹ 10000.00 on planting material, ₹ 2000.00 on irrigation, ₹ 2500.00 on farm yard manure, ₹ 96.50 on fertilizers, ₹ 825.00 on plant protection chemicals, and ₹ 600.00 on training whereas fixed costs were equal to ₹ 6720.00 per hectare with ₹ 1000.00 incurred on depreciation of implements and ₹ 5000.00 as earned value of rented land (EVRL). The per hectare interest on working and fixed capital was worked out to be ₹ 3326.58 and ₹ 720.00, respectively.

Operational costs

The per hectare operational costs consists of yearly expenses on the maintenance of bearing orchard. The item wise and concept wise operational costs for walnut orchards are presented in table 45. The per hectare variable costs and fixed were found to be ₹ 25473.56 and ₹ 6720.00, respectively with a total of ₹ 32193.56 per hectare. The table further revealed that among the working costs, the highest expenditures of ₹ 6994.25.00 per hectare were incurred on fertilizers and lowest of ₹ 1500.00 on intercultural operations.The fixed costs included ₹ 1000.00 on depreciation on machinery and farm inventory, ₹ 5000.00 as earned value of rented land and ₹ 720.00 as interest on fixed capital

Year wise cost and return analysis

The per hectare per year cost and returns from walnut orchards worked out for different age groups are presented in Table 46. The returns starts from 9th year of the establishment of the orchard with average production of 40.00 quintals per hectare during the year 6th to 10th which increases to 75.00 quintals per hectare from 15th year and onwards. As far as the net returns are concerned, it was found to be ₹ 211024.54 per hectare during initial phase of returns which increases to ₹ 417806.44 per hectare during 15th year of the establishment of the orchard and onwards.The benefit cost ratio of 1:8.28 was found during 10th

year which increases to 1:13.98 during 15th year of the establishment of the orchard which proves the viability of the orchard.

Table 44 Operation wise first year establishment costs of walnut orchards with spacing = 10m x 10m (` /ha) No. of Plants per hectare = 100

S.No.	Item	Cost (in ₹)
1	Preparation of land	5000.00
2	Digging, filling and planting	6000.00
3	Planting material/ Plantation cost	10000.00
4	Irrigation	2000.00
5	Farm Yard Manure	2500.00
6	Fertilizers (Urea, DAP, MoP)	96.50
7	Plant Protection Chemicals	825.00
8	Intercultural operations (weeding, hoeing etc)	700.00
9	Training	600.00
10	Interest on working capital @12 per cent per annum	3326.58
A	Total variable cost (1-10)	31048.08
11	Implements/ Depreciation	1000.00
12	Earned value of rented land	5000.00
13	Interest on fixed capital @ 12 per cent per annum	720.00
B	Total fixed cost (11-13)	6720.00
14	Total establishment cost (A+B)	37768.08

Table 45 Item wise and concept wise operational costs of walnut orchards with spacing = 10m x 10m (` /ha)

S.No.	Item	Cost (in ₹)
1	Labour	6000.00
2	Intercultural operations and hoeing, weeding	1500.00
3	Irrigation	2000.00
4	Training/ pruning	2250.00
5	Farm Yard Manure	2500.00
6	Fertilizers (Urea, DAP, MoP)	6994.25
7	Plant protection	1500.00
8	Interest on working capital	2729.31
A	Total variable cost (1-8)	25473.56
9	Depreciation	1000.00
10	EVRL*	50000.00
11	Interest on fixed capital	720.00
B	Total fixed cost (9-11)	6720.00
12	Total operational/ maintenance cost (A+B)	32193.56

* Earned value on rented land

Table 46 Year wise cost and return analysis of walnut orchards with spacing= 10 x 10m (₹/ha)

S.No.	Particulars	Years							
		1	**2**	**3**	**4**	**5**	**Average 6th to 10th year**	**Average 11th-15th year**	**15th year and onwards**
1.	Total cost	37768.08	20968.08	22084.00	23575.50	24365.80	28975.46	30432.50	32193.56
2.	Production (qtls)	0.00	0.00	0.00	0.00	0.00	40.00	55.00	75.00
3.	Gross returns (in ₹)	0.00	0.00	0.00	0.00	0.00	240000.00	330000.00	450000.00
4.	Net returns (in ₹)	-37768.08	-20968.08	-22084.00	-23575.50	-24365.80	211024.54	299567.50	417806.44
5.	Benefit cost ratio	0.00	0.00	0.00	0.00	0.00	1:8.28	1:10.84	1:13.98

❐❐❐

References

Acharya, S. S. and Aggarwal, N. L. 2001. *Agricultural Marketing in India*. Third edition, Oxford & IBH Publishing Company, New Delhi.

Ajani, O. I. 2005. Economic analysis of the marketing of fruit in Lagos State of Nigeria (A case study of Oyingbo, Oshodi and Ikotun markets). *Nigerian Journal of Horticultural Science,* **10**: 38-46.

Bakshi, P., Kour, K. and Jasrotia, A. 2019. Fundamentals of Horticulture. Narendra Publishing House, New Delhi.

Bal, J.S. 2007. Fruit Growing. Kalyani Publishers, New Delhi.

Bhat, Anil and Kachroo, J. 2011. Economics of Production and Marketing of Citrus. Book Published by Daya Publishing House, New Delhi.

Bhat, Anil and Singh, S. P. 2016. Agricultural Marketing: Perspectives and Potentials. Book Published by New India Publishing House.

Chadha, K.L. 2003. Handbook of Horticulture. Indian Council of Agricultural Research (ICAR) pp. 1069.

Iqbal, Mudasir. 2009. *Investment appraisal of mango and ber fruit production in Jammu district of J&K state.* M.Sc. thesis. Sher-e-Kashmir University of Agricultural Sciences and Technology of Jammu, Jammu, India.

Kumar, S., Karol, A., Singh, R. and Vaidya, C. S. 2007. Cost and return from apple cultivation: A study in Himachal Pradesh. *Agricultural Situation in India*, **64**(1): 307-313.

Package and Practices for horticulture crops.2015. Directorate of Extension Sher-e-kashmir University of Agricultural Sciences & Technology of Jammu (SKUAST-J). pp. 147

Package of Practices for cultivation of Fruits. 2008. Punjab Agricultural University (PAU) Ludhiana. pp 127.

Singh, S.P. 1995. Commercial Fruits. Kalyani Publishers, Ludhiana.

Singh, A. 2003. Fruit Physiology and Production. Kalyani Publishers, New Delhi. Pp. 554.

Temperate fruits: Package of Practices. 2011. Directorate of Extension Education, SKUAST-K, Shalimar, Srinagar.pp.123

FSC
www.fsc.org
MIX
Papier | Fördert gute Waldnutzung
FSC® C083411

Zeitfracht Medien GmbH
Ferdinand-Jühlke-Straße 7
99095 Erfurt, Deutschland
produktsicherheit@kolibri360.de